Lecture Notes in Mathematics 1615

Springer
Berlin
Heidelberg
New York
Barcelona
Budapest
Hong Kong
London
Milan
Paris
Santa Clara
Singapore
Tokyo

David B. Massey

Lê Cycles and Hypersurface Singularities

Springer

Author

David B. Massey
Department of Mathematics
Northeastern University
Boston, MA 02115, USA

Cataloging-in-Publication data applied for

Library of Congress Cataloging-in-Publication Data

Massey, David B., 1959-
Lê cycles and hypersurface singularities / David B. Massey.
p. cm. -- (Lecture notes in mathematics ; 1615)
Includes bibliographical references and index.
ISBN 3-540-60395-6 (softcover)
1. Stratified sets. 2. Hypersurfaces. 3. Functions of several complex variables. 4. Singularities (Mathematics) I. Title.
II. Series: Lecture notes in mathematics (Springer-Verlag) ; 1615.
QA3.L28 no. 1615
[QA614.42]
510 s--dc20
[516.3'6] 95-39592
CIP

Mathematics Subject Classification (1991): 32C18, 32B10, 32B15, 32C25

ISBN 3-540-60395-6 Springer-Verlag Berlin Heidelberg New York

Printed in Germany

Typesetting: Camera-ready TEX output by the author
SPIN: 10517520 46/3142-543210 - Printed on acid-free paper

For my father, Robert Brian Massey, on the occasion of his 60th birthday.

PREFACE

In 1976, the result of Lê and Ramanujam [**L-R**] appeared. This result – that the constancy of the Milnor number in a one-parameter family of isolated hypersurface singularities implies topological constancy – is an outstandingly beautiful blend of algebraic geometry and topology.

In 1983, my dissertation advisor, Bill Pardon, gave me three problems to choose from for my thesis: the first was to develop stratified Smith theory, the second I cannot seem to remember, and the third was to generalize the result of Lê and Ramanujam to families of hypersurfaces with one-dimensional singular sets. The third problem was both my favorite and Bill's, and so the topic of my dissertation was decided.

While the results of my dissertation (see [**Mas1**] and [**Mas2**]) were moderately interesting, there were several unsatisfying aspects of my work. The numerical data that I required to be constant in a family seemed too strong and did not seem to generalize easily to higher-dimensional singularities. Moreover, in [**Va1**], Vannier gave a number of stronger results (using completely different techniques).

In 1987, during my second year in a visiting position at the University of Notre Dame, I "fixed" these problems – I developed the Lê cycles and Lê numbers of a hypersurface singularity of arbitrary dimension. The Lê numbers are a generalization of the Milnor number and, by using them, I have been able to generalize a number of results on isolated hypersurface singularities to non-isolated hypersurface singularities. Some of these results appear in [**Mas2**], [**Mas3**], [**Mas5**], and [**M-S**]. However, I have now compiled a number of new results and some interesting examples.

These notes summarize essentially all of the results that I know of concerning the Lê numbers. The types of problems that I can attack with Lê cycles and Lê numbers are problems concerning the Milnor fibration and problems concerning how limiting tangent planes to level hypersurfaces approach a given hypersurface. Some of the results here have appeared elsewhere, but I include the proofs for completeness and because there have been some slight, but important, improvements in many of the statements.

However, many of the results of this work have never appeared; new results include an application of a lemma of Lazarsfeld in Proposition 1.27, an application of Siersma's result on isolated line singularities in Corollary 3.3, a Plücker formula for the Lê numbers in Corollary 4.6, the lexigraphic uppersemicontinuity of the tuple of Lê numbers given in Corollary 4.14, the results on hyperplane arrangements given in Chapter 5, the multi-parameter and nonparametrized versions of results on Thom's a_f condition given in Chapter 6, all the results on aligned singularities given in Chapter 7, and the new characterizations of the Lê cycles given in Chapter 10.

In addition to giving these new results, a major purpose of these notes is to provide a large number of interesting examples. The examples include one in which the Lê cycles move with the coordinate choice (2.3), the coordinate hyperplanes in affine space (2.6), non-reduced plane curves (2.7), quasi-homogeneous singularities (4.7 and 4.8), the specific example of the swallowtail singularity (4.9 and 4.10), and central hyperplane arrangements (5.2).

I believe a word or two is in order about how I chose the terminology *Lê cycles* and *Lê numbers*. I originally was calling these objects *invariant cycles* and *invariant numbers*, but then I found that they were not nearly so "invariant" as I first believed. At that point, I sat there not writing another word for fifteen minutes because I could not think of what else to call these mathematical devices. Finally, I decided to name them after the man whose work had influenced me far more than anyone else's had influenced me up to that point in my career. When I first showed Lê Dũng Tráng my work on the Lê cycles and Lê numbers, it was with a great deal of trepidation. I informed Tráng that if he did not like the work, I would certainly change the names of these cycles and numbers; I am pleased to say that I have not had to alter my terminology.

As this work is a summary of my study of Lê cycles and Lê numbers from 1987 to the present, there are many people that I should thank:

Ambar Chowdhury for his encouragement and many helpful "conversations" at The Commons; Mark Goresky, Lê Dũng Tráng, and Terry Gaffney for bringing me to Northeastern University and for many, many invaluable discussions; a special thanks to Terry for Theorem 10.17; Dirk Siersma for the bounds which appear in the general Lê-Iomdine formulas; Dan Cohen for his help with hyperplane arrangements and for leading me to the chain complex which appears in Theorem 10.9; Dominic Welsh and Günter Ziegler for pointing out the identity between Lê numbers and the Möbius function given in Theorem 5.6; Steve Kleiman for conversations on the relationship between the Lê cycles and the work of Vogel and van Gastel, especially how it implies the Plücker formula of Corollary 4.6; Bob MacPherson for a number of conversations about the derived category and perverse sheaves; Roberto Callejas-Bedregal for one long, bewildering discussion about the Lê cycles, which led to Theorem 10.14; Alex Suciu for the homology groups of the Milnor fibre of the swallowtail in Remark 4.10, and for much encouragement over many cups of coffee; and Robert Gassler, Mike Green, Alex Suciu and, of course, Lê Dũng Tráng for making many helpful suggestions and for spotting many typographical errors in the preliminary versions of this manuscript.

Also, there are some institutions that I must thank:

the Mathematics departments of The University of Notre Dame and Northeastern University, for having me as a visiting professor and treating me so well;

the National Science Foundation, for three years as a post-doctoral research fellow and then for summer support; and The University of Paris VII, for two one-month visits during which much was accomplished.

David B. Massey
Boston, MA
April 24, 1995

TABLE OF CONTENTS

INTRODUCTION

The Lê numbers and Lê cycles generalize the data given by the Milnor number of an isolated hypersurface singularity. In this introduction, we wish to quickly review why the Milnor number of an isolated hypersurface singularity is important. We will then give some previously known general results on non-isolated hypersurface singularities, and indicate the types of results that can be obtained by the machinery contained in the rest of this book.

Let $\mathcal{U}$ be an open neighborhood of the origin in $\mathbb{C}^{n+1}$ and let $f : (\mathcal{U}, 0) \to (\mathbb{C}, 0)$ be an analytic function. Then, the Milnor fibration [**Mi3**], [**Lê5**], [**Ra**] of f at the origin is an object of primary importance in the study of the local, ambient topology of the hypersurface, $V(f) := f^{-1}(0)$, defined by f at the origin. Milnor defined his fibration on a sphere of radius ϵ; however, his Theorem 5.11 of [**Mi3**] leads one to consider a more convenient, equivalent, fibration which lives inside the open ball of radius ϵ. Hence, throughout these notes, we will use the Milnor fibration as defined below.

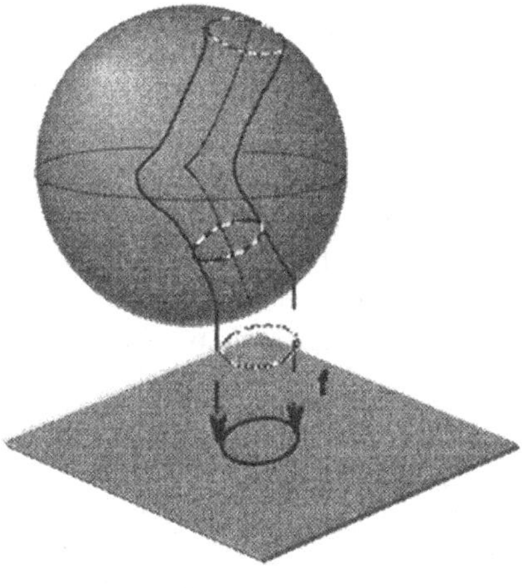

Figure 0.1. The Milnor Fibration inside a ball

For all $\epsilon > 0$, let $\overset{\circ}{B}_\epsilon$ denote the open ball of radius ϵ centered at the origin in $\mathbb{C}^{n+1}$. For all $\eta > 0$, let $\mathbb{D}_\eta$ denote the closed disc centered at the origin in $\mathbb{C}$, and let $\partial\mathbb{D}_\eta$ denote its boundary, which is a circle of radius η. Then, having fixed an analytic function, f, there exists $\epsilon_0 > 0$ such that, for all ϵ such that $0 < \epsilon \leqslant \epsilon_0$, there exists $\eta_\epsilon > 0$ such that, for all η such that $0 < \eta \leqslant \eta_\epsilon$, the restriction of f to a map $\overset{\circ}{B}_\epsilon \cap f^{-1}(\partial\mathbb{D}_\eta) \to \partial\mathbb{D}_\eta$ is a smooth, locally trivial fibration whose diffeomorphism-type is independent of the choice of ϵ and η.

This fibration is called the *Milnor fibration* of f at the origin and the fibre is the *Milnor fibre* of f at the origin, which we denote by $F_{f,\mathbf{0}}$. Hence, the Milnor fibre is a smooth complex n-manifold (of real dimension $2n$). The homotopy-type of the Milnor fibre is an invariant of the local, ambient topological-type of the hypersurface at the origin.

The Results of Milnor

We keep the notation from above; in particular, $\mathcal{U}$ is an open neighborhood of the origin in $\mathbb{C}^{n+1}$ and $f : (\mathcal{U}, \mathbf{0}) \to (\mathbb{C}, 0)$ is an analytic function (actually, for Milnor, f was required to be a polynomial). We will use Σf to denote the critical locus of the map f.

In [**Mi3**], Milnor proved the existence of the object that is now called the Milnor fibration. He also proved that the Milnor fibre, $F_{f,\mathbf{0}}$, has the homotopy-type of a finite n-dimensional CW-complex ([**Mi3**], Theorem 5.1). This implies that all of the homology groups are finitely-generated, are zero above degree n, and that $H_n(F_{f,\mathbf{0}})$ is free Abelian.

In addition, Milnor proved that if f has an isolated critical point at the origin, i.e. $\dim_{\mathbf{0}}\Sigma f = 0$, then $F_{f,\mathbf{0}}$ is $(n-1)$-connected ([**Mi3**], Lemma 6.4). Combining this with the previous result, it follows ([**Mi3**], Theorem 6.5) that, in the case of an isolated singularity, the Milnor fibre has the homotopy-type of a finite bouquet (one-point union) of n-spheres; the number of spheres in this bouquet is the *Milnor number* and is denoted by μ (or $\mu_{\mathbf{0}}(f)$, or some other such variant). In particular, the reduced homology is trivial except in degree n, and there the homology group is $\mathbb{Z}^{\mu}$. The Milnor number can be calculated algebraically by taking the dimension as a complex vector space of the algebra $\mathcal{O}_{\mathbf{0}}^{n+1}/J(f)$, where $\mathcal{O}_{\mathbf{0}}^{n+1}$ denotes the ring of analytic germs at the origin and $J(f)$ denotes the Jacobian ideal $\langle \frac{\partial f}{\partial z_0}, \dots, \frac{\partial f}{\partial z_n} \rangle$.

So that we can do an example, there is one final result of Milnor's that we wish to mention here. Suppose that f is a weighted homogeneous polynomial (i.e. there exist positive integers $r_0, \dots, r_n$ such that $f(z_0{}^{r_0}, \dots, z_n{}^{r_n})$ is a homogeneous polynomial). Then, ([**Mi3**], Lemma 9.4) the Milnor fibre, $F_{f,\mathbf{0}}$, is diffeomorphic to $f^{-1}(1)$.

Example 0.2. As an example, consider $f = xyz$, which defines a hypersurface in $\mathbb{C}^3$ consisting of the three coordinate planes. Thus, $V(f)$ is a hypersurface with a one-dimensional singular set consisting of the three coordinate axes.

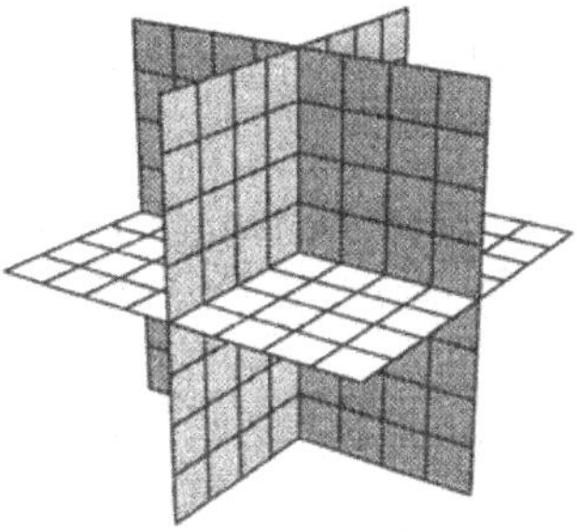

Figure 0.3. The coordinate hyperplanes

By the above result on weighted homogeneous polynomials, the Milnor fibre is diffeomorphic to the set of points where $xyz = 1$; but, this is where $x \neq 0$, $y \neq 0$, and $x = \frac{1}{yz}$.

Thus, $F_{f,0} \cong \mathbb{C}^* \times \mathbb{C}^*$, where $\mathbb{C}^* = \mathbb{C} - 0$. In particular, $F_{f,0}$ is homotopy-equivalent to the product of two circles, and so has non-zero homology in degrees 0, 1, and 2.

Further Results

We wish to consider another classic example: the Whitney umbrella.

Example 0.4. The Whitney umbrella is the hypersurface in $\mathbb{C}^3$ defined by the vanishing of $f = y^2 - zx^2$.

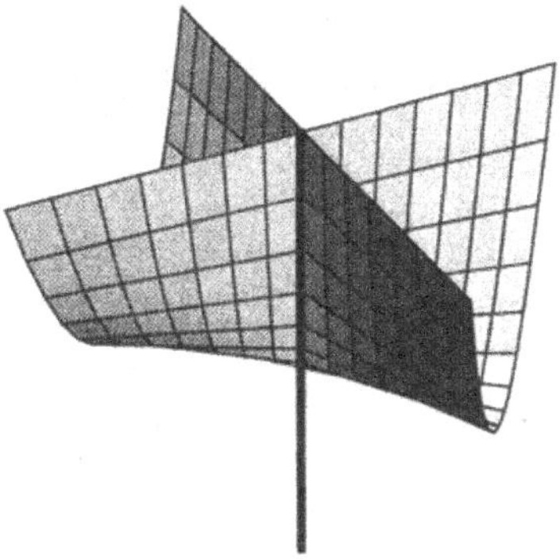

Figure 0.5. The Whitney umbrella

Here, we have drawn the picture over the real numbers – this is the rarely-seen picture that explains the word "umbrella" in the name of this example. The "handle" of this umbrella is not usually drawn when one is in the complex setting, for the inclusion of this line gives the impression that the local dimension of the hypersurface is not constant; something which is not possible over the complex numbers. A second reason why one rarely sees the above picture is that one frequently encounters the Whitney umbrella as a family of nodes degenerating to a cusp; this representation is achieved by making the analytic change of coordinates $z = x + t$ to obtain $f = y^2 - x^3 - tx^2$ (see Example 1.12).

To determine the homotopy-type of the Milnor fibre of the Whitney umbrella at the origin, we need a new result.

The result we need is that if we have an analytic function $g(z_0, \dots, z_n)$ and a variable y, disjoint from the z's, then the Milnor fibre of $y^2 + g(z_0, \dots, z_n)$ is homotopy-equivalent to the suspension of the Milnor fibre of g. By an abuse of language, one frequently says that the singularity of $y^2 + g(z_0, \dots, z_n)$ is the *suspension of the singularity* of g.

So, in our example, $F_{y^2-zx^2,\mathbf{0}}$ is homotopy-equivalent to the suspension of $F_{zx^2,\mathbf{0}}$. But, as zx^2 is homogeneous,

$$F_{zx^2,\mathbf{0}} \cong \left\{(z,x) \mid zx^2 = 1\right\} = \left\{\left(\frac{1}{x^2}, x\right) \,\middle|\, x \neq 0\right\} \cong \mathbb{C}^*.$$

Thus, $F_{y^2-zx^2,\mathbf{0}}$ is homotopy equivalent to the suspension of a circle, i.e. the Milnor fibre of f at the origin is homotopy-equivalent to a 2-sphere.

The suspension result used above is a special case of a much more general result proved in various forms in **[Ok]**, **[Sak]**, and **[Se-Th]**. This result states that:

Theorem 0.6. If $f : (\mathcal{U}, \mathbf{0}) \to (\mathbb{C}, 0)$ and $g : (\mathcal{U}', \mathbf{0}) \to (\mathbb{C}, 0)$ are analytic functions, then the Milnor fibre of the function $h : (\mathcal{U} \times \mathcal{U}', \mathbf{0}) \to (\mathbb{C}, 0)$ defined by $h(\mathbf{w}, \mathbf{z}) := f(\mathbf{w}) + g(\mathbf{z})$ is homotopy-equivalent to the join, $F_{f,\mathbf{0}} * F_{g,\mathbf{0}}$, of the Milnor fibres of f and g.

This determines the homology of $F_{h,\mathbf{0}}$ in a simple way, since the reduced homology of the join of two spaces X and Y is given by

$$\widetilde{H}_{j+1}(X * Y) = \sum_{k+l=j} \widetilde{H}_k(X) \otimes \widetilde{H}_l(Y) \oplus \sum_{k+l=j-1} \operatorname{Tor}\left(\widetilde{H}_k(X), \widetilde{H}_l(Y)\right).$$

Returning now to Example 0.2 where $f = xyz$, we see that $F_{f,\mathbf{0}}$ need not have the homotopy-type of a bouquet of spheres when the singularity is non-isolated. However, there is the general result of Kato and Matsumoto **[K-M]**:

Theorem 0.7. *If $s := \dim_0 \Sigma f$, then $F_{f,0}$ is $(n-s-1)$-connected; in particular, when $s = 0$, we recover the result of Milnor.*

Moreover, this is the best possible general bound on the connectivity of the Milnor fibre, as is shown by:

Example 0.8. Consider

$$g := z_0 z_1 \dots z_{s+1} + z_{s+2}^2 + \cdots + z_n^2;$$

we leave it as an exercise for the reader to verify, using our earlier methods, that this g has an s-dimensional critical locus at the origin and $F_{g,0}$ has non-trivial homology in dimension $n - s$.

Lê's Attaching Result

The result of Kato and Matsumoto can be obtained from a more general result of Lê; a result which is one of few which allows calculations concerning the homology of the Milnor fibre for an arbitrary hypersurface singularity.

Let $\mathcal{U}$ be an open neighborhood of the origin in $\mathbb{C}^{n+1}$ and let $f : (\mathcal{U}, 0) \to (\mathbb{C}, 0)$ be an analytic function. Let $L : \mathbb{C}^{n+1} \to \mathbb{C}$ be a generic linear form. Then, it is easy to see that if $\dim_0 \Sigma f \geqslant 1$, then $\dim_0 \Sigma(f_{|_{V(L)}}) = (\dim_0 \Sigma f) - 1$.

Now, the main result of [**Lê1**] is:

Theorem 0.9. *The Milnor fibre $F_{f,0}$ is obtained from the Milnor fibre $F_{f_{|_{V(L)}},0}$ by attaching a certain number of n-handles (n-cells on the homotopy level); this number of attached n-handles is given by the intersection number $\left(\Gamma^1_{f,L} \cdot V(f)\right)_0$, where $\Gamma^1_{f,L}$ denotes the relative polar curve of f with respect to L.*

We will define the polar curve and discuss how to calculate intersection numbers in Chapter 1, but we can already see that Kato and Matsumoto's result follows inductively from Theorem 0.9 since we already know Milnor's result for isolated singularities and because attaching handles of index k does not affect the connectivity in dimensions $\leqslant k - 2$.

Not only does Lê's result imply Kato and Matsumoto's, but – assuming that $\left(\Gamma^1_{f,L} \cdot V(f)\right)_0$ is effectively calculable – Lê's result enables the calculation of the Euler characteristic of the Milnor fibre, together with some Morse-type inequalities on the Betti numbers of the Milnor fibre; for instance, the n-th Betti number, $b_n(F_{f,0})$, must be less than or equal to $\left(\Gamma^1_{f,L} \cdot V(f)\right)_0$.

Unfortunately, the Morse inequalities above are usually far from being equalities. Of course, the real value of Lê's result is that it allows one to calculate some important information even in the cases where the homotopy-type of the Milnor fibre cannot be determined by other means.

The Result of Lê and Ramanujam

As the homotopy-type of the Milnor fibre is an invariant of the local, ambient topological-type of the hypersurface at the origin, if one has a family of hypersurfaces with isolated singularities in which the local, ambient, topological-type is constant, then the Milnor number must remain constant in the family. In 1976, Lê and Ramanujam [**L-R**] proved the converse of this; we describe their result now.

Let $\overset{\circ}{\mathbb{D}}$ be an open disc about the origin in $\mathbb{C}$, let $\mathcal{U}$ be an open neighborhood of the origin in $\mathbb{C}^{n+1}$, and let $f : (\overset{\circ}{\mathbb{D}} \times \mathcal{U}, \overset{\circ}{\mathbb{D}} \times \mathbf{0}) \to (\mathbb{C}, 0)$ be an analytic function; we write f_t for the function defined by $f_t(\mathbf{z}) := f(t, \mathbf{z})$. Lê and Ramanujam proved:

Theorem 0.10. *Suppose that, for all small t,* $\dim_{\mathbf{0}} \Sigma f_t = 0$ *and that the Milnor number of f_t at the origin is independent of t. Then, for all small t,*

i) the fibre-homotopy type of the Milnor fibrations of f_t at the origin is independent of t;

and, if $n \neq 2$,

ii) the diffeomorphism-type of the Milnor fibrations of f_t and the local, ambient, topological-type of $V(f_t)$ at the origin are independent of t.

The Result of Lê and Saito

The result of Lê and Saito again deals with families of singularities, so we continue with $f : (\overset{\circ}{\mathbb{D}} \times \mathcal{U}, \overset{\circ}{\mathbb{D}} \times \mathbf{0}) \to (\mathbb{C}, 0)$ as above. The result of [**L-S**] tells one how limiting tangent spaces to nearby level hypersurfaces of f approach the singularity.

Theorem 0.11. *Suppose that, for all small t,* $\dim_{\mathbf{0}} \Sigma f_t = 0$ *and that the Milnor number of f_t at the origin is independent of t. Then, $\overset{\circ}{\mathbb{D}} \times \mathbf{0}$ satisfies Thom's a_f condition at the origin with respect to the ambient stratum, i.e. if $\mathbf{p}_i$ is a sequence of points in $\overset{\circ}{\mathbb{D}} \times \mathcal{U} - \Sigma f$ such that $\mathbf{p}_i \to \mathbf{0}$ and such that $T_{\mathbf{p}_i} V(f - f(\mathbf{p}_i))$ converges* *to some* $\mathcal{T}$, *then* $\mathbb{C} \times \mathbf{0} = \mathcal{T}_{\mathbf{0}}(\overset{\circ}{\mathbb{D}} \times \mathbf{0}) \subseteq \mathcal{T}$.

Generalizing the Milnor Number

So, suppose we have a single analytic function, $f : (\mathcal{U}, \mathbf{0}) \to (\mathbb{C}, 0)$ with a critical locus of arbitrary dimension $s := \dim_{\mathbf{0}} \Sigma f$. What properties would we want generalized Milnor numbers of f at $\mathbf{0}$ to have?

First, associated to f, we want there to be $s+1$ numbers which are effectively calculable; call the numbers $\lambda_f^0, \ldots, \lambda_f^s$. In the case of an isolated singularity, we want λ_f^0 to be the Milnor number of f and all other λ_f^i to be zero.

For arbitrary s, we would like to generalize Milnor's result for isolated singularities and show that the Milnor fibre of f at the origin has a handle decomposition in which the number of attached handles of each index are given by the appropriate λ_f^i.

Finally, we would like to have generalizations of the results of Lê and Ramanujam and Lê and Saito to families of hypersurface singularities of arbitrary dimension.

As we shall see, the Lê numbers succeed at these goals to a great degree.

Chapter 1. DEFINITIONS AND BASIC PROPERTIES

In this chapter, we define and prove some elementary results about the fundamental objects of study in these notes – the Lê cycles and Lê numbers. The Lê cycles are analytic cycles which, in a sense, decompose the critical locus of an analytic function. The Lê numbers are intersection numbers of the Lê cycles with certain linear subspaces.

To define the Lê cycles, we first need to define the relative polar cycles, which are the cycles associated to the relative polar varieties. The relative polar varieties were studied by Lê and Teissier in a number of places (see, for instance, [**L-T2**], [**Te2**], and [**Te3**]). Lê and Teissier define the relative polar varieties of a function with respect to generic linear flags, and they usually assume that the flags have been chosen generically enough so that the relative polar varieties have many special properties.

However, the whole theory seems to behave more nicely if one does not require the flags to be quite so generic, and then works with possibly non-reduced schemes and cycles. Therefore, after we fix some notation and terminology, we begin by discussing the algebraic operations that we will use to obtain the non-reduced relative polar varieties.

Throughout this chapter, $\mathcal{U}$ will denote an open subset of $\mathbb{C}^{n+1}$ containing the origin and $h : (\mathcal{U}, \mathbf{0}) \to (\mathbb{C}, 0)$ will be an analytic function.

We wish to consider schemes, cycles, and sets. If h is, in fact, a polynomial, then we may use algebraic schemes, cycles, and sets. However, as we wish to treat the more general analytic case, we should clarify what we mean by the terms scheme and cycle.

In the analytic setting, by *scheme*, we actually mean a (not necessarily reduced) complex analytic space, $(X, \mathcal{O}_X)$, in the sense of [**G-R1**] and [**G-R2**]. By the irreducible components of X, we mean simply the irreducible components of the underlying analytic set X. If we concentrate our attention on the germ of X at some point $\mathbf{p}$, then we may discuss embedded and non-embedded (a.k.a. isolated) components of the germ of X at $\mathbf{p}$ – these correspond to non-minimal and minimal primes, respectively, in the set of associated primes of the Noetherian local ring $\mathcal{O}_{X,\mathbf{p}}$.

If $\mathcal{U}$ is an open subset of affine space and α is a coherent sheaf of ideals in $\mathcal{O}_{\mathcal{U}}$, then we write $V(\alpha)$ for the possibly non-reduced analytic space defined by the vanishing of α.

Given an analytic space X (with its reduced structure), an *analytic cycle* in X is a formal sum $\sum m_V[V]$, where the V's are irreducible analytic subsets of X, the m_V's are integers, and the collection $\{V\}$ is a locally finite collection of subsets of X. As a cycle is a locally finite sum, and as we will normally be concentrating on the germ of an analytic space at a point, usually we can safely assume that a cycle is actually a finite formal sum.

Given an analytic space, $(X, \mathcal{O}_X)$, we wish to define the cycle associated to $(X, \mathcal{O}_X)$. In the algebraic context, this is given by Fulton in [**Fu**, 1.5] as

$$[X] := \sum m_V [V],$$

where the V's run over all the irreducible components of X, and m_V equals the length of the ring $\mathcal{O}_{X,V}$, the local ring of X along V. In the analytic context, we wish to use the same definition, but we must be more careful in defining the m_V.

Define m_V as follows. Take a point $\mathbf{p}$ in V. The germ of V at $\mathbf{p}$ breaks up into irreducible germ components $(V_{\mathbf{p}})_i$. Take any one of the $(V_{\mathbf{p}})_i$ and let m_V equal the length of the ring $(\mathcal{O}_{X,\mathbf{p}})_{(V_{\mathbf{p}})_i}$ (that is, the local ring of X at $\mathbf{p}$ localized at the prime corresponding to $(V_{\mathbf{p}})_i$). This number is independent of the point $\mathbf{p}$ in V and the choice of $(V_{\mathbf{p}})_i$.

With this notation, it is fundamental that, if $f, g \in \mathcal{O}_{\mathcal{U}}$, then $[V(fg)] = [V(f)] + [V(g)]$; in particular, $[V(f^m)] = m[V(f)]$.

For clarification of what structure we are considering, we shall at times enclose cycles in square brackets, [ ], and analytic sets in a pair of vertical lines,| |. Occasionally, when the notation becomes cumbersome, we shall simply state explicitly whether we are considering V as a scheme, a cycle, or a set.

By the intersection of a collection of closed subschemes, we mean the scheme defined by the sum of the underlying ideal sheaves. By the union of a finite collection of closed subschemes, we mean the scheme defined by the intersection (not the product) of the underlying ideal sheaves. We say that two subschemes, V and W, are equal up to embedded component provided that, in each stalk, the isolated components of the defining ideals (those corresponding to minimal primes) are equal. Our main concern with this last notion is that it implies that the cycles [V] and [W] are equal. We say that two cycles are equal at a point, $\mathbf{p}$, provided that the portions of each cycle which pass through $\mathbf{p}$ are equal. Finally, when we say that a space, X, is purely k-dimensional at a point, $\mathbf{p}$, we mean to allow for the vacuous case where X has no components through $\mathbf{p}$.

If we have two irreducible subschemes V and W in an open subset $\mathcal{U}$ of some affine space, V and W are said to *intersect properly* in $\mathcal{U}$ provided that $\operatorname{codim} V \cap W = \operatorname{codim} V + \operatorname{codim} W$; when this is the case, the *intersection product* of $[V]$ and $[W]$ is defined by $[V] \cdot [W] = [V \cap W]$. Two cycles $\sum m_i [V_i]$ and $\sum n_j [W_j]$ are said to intersect properly if V_i and W_j intersect properly for all i and j; when this is the case, the intersection product is extended bilinearly by defining

$$\sum m_i [V_i] \cdot \sum n_j [W_j] = \sum m_i n_j \left([V_i] \cdot [W_j]\right) = \sum m_i n_j \left([V_i \cap W_j]\right).$$

If two cycles C_1 and C_2 intersect properly and $C_1 \cdot C_2 = \sum p_k [Z_k]$, where the Z_k are irreducible, then the *intersection number* of C_1 and C_2 at Z_k, $(C_1 \cdot C_2)_{Z_k}$, is defined to be p_k; that is, the number of times Z_k occurs in the intersection, counted with multiplicity. Throughout these notes, we use intersection numbers only when C_1 and C_2 have complementary codimensions; in this case, all the Z_k are merely points.

Finally, given a point $\mathbf{p} \in \mathcal{U}$, a curve W in $\mathcal{U}$ which is irreducible at $\mathbf{p}$, and a hypersurface $V(f) \subseteq \mathcal{U}$ which intersects W properly at $\mathbf{p}$, there is a very useful way to calculate the intersection number $([W] \cdot [V(f)])_{\mathbf{p}}$. One takes a local parameterization $\phi(t)$ of W which takes 0 to $\mathbf{p}$, and then $([W] \cdot [V(f)])_{\mathbf{p}} = \text{mult}_t f(\phi(t)) =$ the degree of the lowest non-zero term.

We now wish to define the algebraic device which we use to define the relative polar varieties as possibly non-reduced schemes.

Let W be analytic subset of some open subset $\mathcal{U}$ in some affine space and let α be a coherent sheaf of ideals in $\mathcal{O}_{\mathcal{U}}$. At each point, $\mathbf{x}$, of $V(\alpha)$, we wish to consider scheme-theoretically those components of $V(\alpha)$ which are not contained in $|W|$.

Definition 1.1. Let A denote $\mathcal{O}_{\mathcal{U},\mathbf{x}}$; we write $\alpha_{\mathbf{x}}$ for the stalk of α in A. Let S be the multiplicatively closed set $A - \bigcup p$ where the union is over all $p \in \text{Ass}(A/\alpha_{\mathbf{x}})$ with $|V(p)| \not\subseteq |W|$. Then, we define $\alpha_{\mathbf{x}}/W$ to equal $S^{-1}\alpha_{\mathbf{x}} \cap A$. Thus, $\alpha_{\mathbf{x}}/W$ is the ideal in A consisting of the intersection of those primary components, q, (possibly embedded), of $\alpha_{\mathbf{x}}$ such that $|V(q)| \not\subseteq |W|$.

Now, we have defined $\alpha_{\mathbf{x}}/W$ in each stalk. By [**Si-Tr**], if we perform this operation simultaneously at all points of $V(\alpha)$, then we obtain a coherent sheaf of ideals called a *gap sheaf*; we write this sheaf as α/W. If $V = V(\alpha)$, we let V/W denote the scheme $V(\alpha/W)$.

It is important to note that the scheme V/W does not depend on the structure of W as a scheme, but only as an analytic set.

The following lemma is very useful for calculating V/W, and is merely an exercise in localization (see [**Mas3**]).

Lemma 1.2. *Let $(X, \mathcal{O}_X)$ be an analytic space, let α, β, and γ be coherent sheaves of ideals in $\mathcal{O}_X$, let $f, g \in \mathcal{O}_X$, and let W be an analytic subset of X. Then,*

i) $(\alpha + \beta)/W = (\alpha/W + \beta)/W$, *and thus, as schemes,*

$$(V(\alpha) \cap V(\beta))/W = (V(\alpha/W) \cap V(\beta))/W;$$

ii) *if* $V(\alpha + \gamma) \subseteq W$, *then* $((\alpha \cap \beta) + \gamma)/W = (\beta + \gamma)/W$, *and thus, as schemes,*

$$((V(\alpha) \cup V(\beta)) \cap V(\gamma))/W = (V(\beta) \cap V(\gamma))/W;$$

iii) *if* $V(\alpha+g) \subseteq W$, *then* $(\alpha+fg)/W = (\alpha+f)/W$, *and thus, as schemes,*

$$(V(\alpha) \cap V(fg))/W = (V(\alpha) \cap V(f))/W.$$

We will now use these gap sheaves to define the relative polar varieties of an analytic function $h : (\mathcal{U}, \mathbf{0}) \to (\mathbb{C}, 0)$, where $\mathcal{U}$ is an open subset of $\mathbb{C}^{n+1}$. The key features of this definition are that the polar varieties are not necessarily reduced and that the dimension of the critical locus of h is allowed to be arbitrary. The reader who is familiar with the works of Lê and Teissier ([**L-T2**], [**Te2**], [**Te3**]) should note that we index by the generic dimension instead of the codimension.

There is one further difference between our presentation of the relative polar varieties and that of Lê and Teissier; instead of fixing a complete flag inside the ambient affine space, we fix a linear choice of coordinates $\mathbf{z} = (z_0, \dots, z_n)$ for $\mathbb{C}^{n+1}$. We do this because we frequently find it useful to have the linear functions $z_0, \dots, z_n$ at our disposal.

Definition 1.3. For $0 \leqslant k \leqslant n$, the *k-th (relative) polar variety*, $\Gamma^k_{h,\mathbf{z}}$, of h with respect to $\mathbf{z}$ is the scheme $V\left(\frac{\partial h}{\partial z_k}, \dots, \frac{\partial h}{\partial z_n}\right) \Big/ \Sigma h$ (see [**Mas2**], [**Mas3**], [**Mas5**]). If the choice of the coordinate system is clear, we will often simply write Γ^k_h.

Thus, on the level of ideals, $\Gamma^k_{h,\mathbf{z}}$ consists of those components of the scheme $V\left(\frac{\partial h}{\partial z_k}, \dots, \frac{\partial h}{\partial z_n}\right)$ which are not contained in $|\Sigma h|$. Note, in particular, that $\Gamma^0_{h,\mathbf{z}}$ is empty.

Naturally, we define *the k-th polar cycle* of h with respect to $\mathbf{z}$ to be the analytic cycle $\left[\Gamma^k_{h,\mathbf{z}}\right]$.

Clearly, as sets, $\emptyset = \Gamma^0_{h,\mathbf{z}} \subseteq \Gamma^1_{h,\mathbf{z}} \subseteq \dots \subseteq \Gamma^{n+1}_{h,\mathbf{z}} = \mathcal{U}$. In fact, by 1.2.i, we have that :

Proposition 1.4. $\left(\Gamma^{k+1}_{h,\mathbf{z}} \cap V\left(\frac{\partial h}{\partial z_k}\right)\right) \Big/ \Sigma h = \Gamma^k_{h,\mathbf{z}}$ *as schemes, and thus all the components of the cycle* $\left[\Gamma^{k+1}_{h,\mathbf{z}} \cap V\left(\frac{\partial h}{\partial z_k}\right)\right] - \left[\Gamma^k_{h,\mathbf{z}}\right]$ *are contained in the critical set of the map* h.

As the ideal $\langle\frac{\partial h}{\partial z_k}, \dots, \frac{\partial h}{\partial z_n}\rangle$ is invariant under any linear change of coordinates which leaves $V(z_0, \dots, z_{k-1})$ invariant, we see that the scheme $\Gamma^k_{h,\mathbf{z}}$ depends only on h and the choice of the first k coordinates. At times, it will be convenient to subscript the k-th polar variety with only the first k coordinates instead of the whole coordinate system; for instance, we write Γ^1_{h,z_0} for the polar curve.

While it is immediate from the number of defining equations that every component of the analytic set $\left|\Gamma^k_{h,\mathbf{z}}\right|$ has dimension at least k, one usually requires that the coordinate system be suitably generic so that the dimension of $\Gamma^k_{h,\mathbf{z}}$ equals k. When this dimension condition is satisfied, we wish to see that $\Gamma^k_{h,\mathbf{z}}$ has no embedded components. For this, we need two easy lemmas.

Lemma 1.5. *Let $\mathbf{p} \in V := V(g_1, \dots, g_d) \subseteq \mathcal{U}$ = an open subset of $\mathbb{C}^{n+1}$, and let W equal the union of the components of V through $\mathbf{p}$ of dimension greater than $n+1-d$. Then, V/W has no embedded subvarieties through $\mathbf{p}$.*

Proof. By definition, V/W contains no components – embedded or isolated – contained in W. Thus, if V/W has an embedded component through $\mathbf{p}$, that component must contain points $\mathbf{q}$, arbitrarily close to $\mathbf{p}$, which are not contained in W. But, the localization of V/W at any point outside of W is clearly Cohen-Macaulay, since at such a point V/W is a local complete intersection. As Cohen-Macaulay rings have no embedded components, this would contradict the existence of an embedded component through $\mathbf{q}$. □

Lemma 1.6. *If $V := V(g_1, \dots, g_d)$ and $\dim_{\mathbf{p}} V/Z = n+1-d$, then V/Z has no embedded components through $\mathbf{p}$.*

Proof. In the notation of Lemma 1.5, this follows from the fact that Z must contain W. □

It follows immediately that we have

Proposition 1.7. *If $\dim_{\mathbf{p}}\Gamma^k_{h,\mathbf{z}} = k$, then $\Gamma^k_{h,\mathbf{z}}$ has no embedded subvarieties through the point $\mathbf{p}$.*

Proposition 1.8. *If $\dim_{\mathbf{p}}\Sigma h < k$, then $\Gamma^k_{h,\mathbf{z}} = V\left(\frac{\partial h}{\partial z_k}, \dots, \frac{\partial h}{\partial z_n}\right)$ as cycles at $\mathbf{p}$. If, in addition, $\dim_{\mathbf{p}}\Gamma^k_{h,\mathbf{z}} = k$, then the equality above holds as schemes at $\mathbf{p}$.*

Proof. As schemes, $V := V\left(\frac{\partial h}{\partial z_k}, \dots, \frac{\partial h}{\partial z_n}\right)$ consists of the components not contained in Σh – these comprise $\Gamma^k_{h,\mathbf{z}}$ – together with those contained in Σh. By the number of defining equations, every isolated component of V must have dimension at least k. Thus, if $\dim_{\mathbf{p}}\Sigma h < k$, the only components of V which are contained in Σh must be embedded. Therefore, $V\left(\frac{\partial h}{\partial z_k}, \dots, \frac{\partial h}{\partial z_n}\right)$ equals $\Gamma^k_{h,\mathbf{z}}$ up to embedded component and, hence, they are equal as cycles.

But this certainly implies that $\Gamma^k_{h,\mathbf{z}}$ and V are equal as germs of sets at $\mathbf{p}$. Thus, if $\dim_{\mathbf{p}}\Gamma^k_{h,\mathbf{z}} = k$, then $\dim_{\mathbf{p}} V = k$, i.e. V is a local complete intersection at $\mathbf{p}$. Hence, V has no embedded components at $\mathbf{p}$. The second statement follows. □

Definition 1.9. If the intersection of $\Gamma^k_{h,\mathbf{z}}$ and $V(z_0 - p_0, \dots, z_{k-1} - p_{k-1})$ is purely zero-dimensional at a point $\mathbf{p} = (p_0, \dots, p_n)$ (i.e. either $\mathbf{p}$ is an isolated point of the intersection or $\mathbf{p}$ is not in the intersection), then we say that the *k-th polar number, $\gamma^k_{h,\mathbf{z}}(\mathbf{p})$, is defined* and we set $\gamma^k_{h,\mathbf{z}}(\mathbf{p})$ equal to the intersection number

$$\left(\Gamma^k_{h,\mathbf{z}} \cdot V(z_0 - p_0, \dots, z_{k-1} - p_{k-1})\right)_{\mathbf{p}} .$$

(We use the term polar **numbers**, instead of polar multiplicities, since we are not assuming that the coordinates are so generic that this intersection number gives the multiplicities.)

Note that, if $\gamma^k_{h,\mathbf{z}}$ is defined at $\mathbf{p}$, then it must be defined at all points near $\mathbf{p}$. Note also that, if $\gamma^k_{h,\mathbf{z}}(\mathbf{p})$ is defined, then $\Gamma^k_{h,\mathbf{z}}$ must be purely k-dimensional at $\mathbf{p}$ and so – by 1.7 – $\Gamma^k_{h,\mathbf{z}}$ has no embedded components at $\mathbf{p}$.

Remark 1.10. As sets,

$$\Sigma(h_{|_{V(z_0-p_0,\dots,z_{k-1}-p_{k-1})}}) = V\left(z_0 - p_0, \dots, z_{k-1} - p_{k-1}, \frac{\partial h}{\partial z_k}, \dots, \frac{\partial h}{\partial z_n}\right)$$
$$= V(z_0 - p_0, \dots, z_{k-1} - p_{k-1}) \cap \left(\Sigma h \cup \Gamma^k_{h,\mathbf{z}}\right).$$

Hence, if $\gamma^k_{h,\mathbf{z}}(\mathbf{p})$ is defined and $\mathbf{p} \in \Sigma h$, then

$$\Sigma(h_{|_{V(z_0-p_0,\dots,z_{k-1}-p_{k-1})}}) = V(z_0 - p_0, \dots, z_{k-1} - p_{k-1}) \cap \Sigma h$$

at $\mathbf{p}$.

We now wish to define the Lê cycles. Unlike the polar varieties and cycles, the Lê cycles are supported on the critical set of h itself. These cycles demonstrate a number of properties which generalize the data given by the Milnor number for an isolated singularity.

Definition 1.11. For $0 \leqslant k \leqslant n$, we define the *k-th* Lê *cycle of h with respect to* **z**, $\left[\Lambda^k_{h,\mathbf{z}}\right]$, to be

$$\left[\Gamma^{k+1}_{h,\mathbf{z}} \cap V\left(\frac{\partial h}{\partial z_k}\right)\right] - \left[\Gamma^k_{h,\mathbf{z}}\right].$$

If the choice of coordinate system is clear, we will sometimes simply write $[\Lambda_h^k]$. Also, as we have given the Lê cycles no structure as schemes, we will usually omit the brackets and write $\Lambda_{h,\mathbf{z}}^k$ to denote the Lê cycle – unless we explicitly state that we are considering it as a set only.

Note that, as every component of $\Gamma_{h,\mathbf{z}}^{k+1}$ has dimension at least $k+1$, every component of $\Lambda_{h,\mathbf{z}}^k$ has dimension at least k. We say that the cycle $\left[\Lambda_{h,\mathbf{z}}^k\right]$ or the set $\left|\Lambda_{h,\mathbf{z}}^k\right|$ has *correct dimension at a point* $\mathbf{p}$ provided that $\left|\Lambda_{h,\mathbf{z}}^k\right|$ is purely k-dimensional at $\mathbf{p}$.

We define the *k-th* Lê *number of h at* $\mathbf{p}$ *with respect to* $\mathbf{z}$, $\lambda_{h,\mathbf{z}}^k(\mathbf{p})$, to equal the intersection number $\left(\Lambda_{h,\mathbf{z}}^k \cdot V(z_0 - p_0, \dots, z_{k-1} - p_{k-1})\right)_{\mathbf{p}}$, provided this intersection is purely zero-dimensional at $\mathbf{p}$. If this intersection is not purely zero-dimensional at $\mathbf{p}$, then we say that the k-th Lê number (of h at $\mathbf{p}$ with respect to $\mathbf{z}$) is *undefined.* Here, when $k=0$, we mean that

$$\lambda_{h,\mathbf{z}}^0(\mathbf{p}) = \left(\Lambda_{h,\mathbf{z}}^0 \cdot \mathcal{U}\right)_{\mathbf{p}} =$$

$$\left[\Gamma_{h,\mathbf{z}}^1 \cap V\left(\frac{\partial h}{\partial z_0}\right)\right]_{\mathbf{p}} = \left(\left[\Gamma_{h,\mathbf{z}}^1\right] \cdot \left[V\left(\frac{\partial h}{\partial z_0}\right)\right]\right)_{\mathbf{p}}.$$

(This last equality holds whenever $\Gamma_{h,\mathbf{z}}^1$ is one-dimensional at $\mathbf{p}$, for then $\Gamma_{h,\mathbf{z}}^1$ has no embedded components by 1.7 and $\Gamma_{h,\mathbf{z}}^1 \cap V\left(\frac{\partial h}{\partial z_0}\right)$ is zero-dimensional. See Lemma 1.17.)

Note that if $\lambda_{h,\mathbf{z}}^k(\mathbf{p})$ is defined, then $\lambda_{h,\mathbf{z}}^k$ is defined at all points near $\mathbf{p}$ and $\left|\Lambda_{h,\mathbf{z}}^k\right|$ must have correct dimension at $\mathbf{p}$. Also note that, since $\Gamma_{h,\mathbf{z}}^{k+1}$ and $\Gamma_{h,\mathbf{z}}^k$ depend only on the choice of the coordinates z_0 through z_k, the k-th Lê cycle, $\left[\Lambda_{h,\mathbf{z}}^k\right]$, depends only on the choice of $(z_0, \dots, z_k)$. Finally, note that if h is a polynomial, then since we are taking **linear** coordinates, we remain inside the algebraic category.

The sum of the Lê cycles exactly corresponds to a Vogel cycle of the Jacobian ideal. See [**Gas1**], [**Gas2**], and [**Vo**].

While we shall defer most of our examples until later – when we will have more results to play with – it is instructive to include at least one at this early stage.

Example 1.12. Let $h = y^2 - x^3 - tx^2$; this is the Whitney umbrella of Example 0.4, but written as a family of nodes degenerating to a cusp. We fix the coordinate system $\mathbf{z} = (t, x, y)$ and will suppress any further reference to it.

We find

$$\Sigma h = V(-x^2,\ -3x^2 - 2tx,\ 2y) = V(x, y).$$

Thus, the critical locus of h is one-dimensional and consists of the t-axis.

Now the critical locus is one-dimensional, while the dimension of every component of $V\left(\frac{\partial h}{\partial y}\right)$ is at least two. Hence, $V\left(\frac{\partial h}{\partial y}\right)$ cannot possibly have any components contained in Σh and, therefore, we begin calculating polar varieties with Γ^2_h. We have simply

$$\Gamma^2_h = V\left(\frac{\partial h}{\partial y}\right) = V(2y) = V(y)$$

with no components to dispose of.

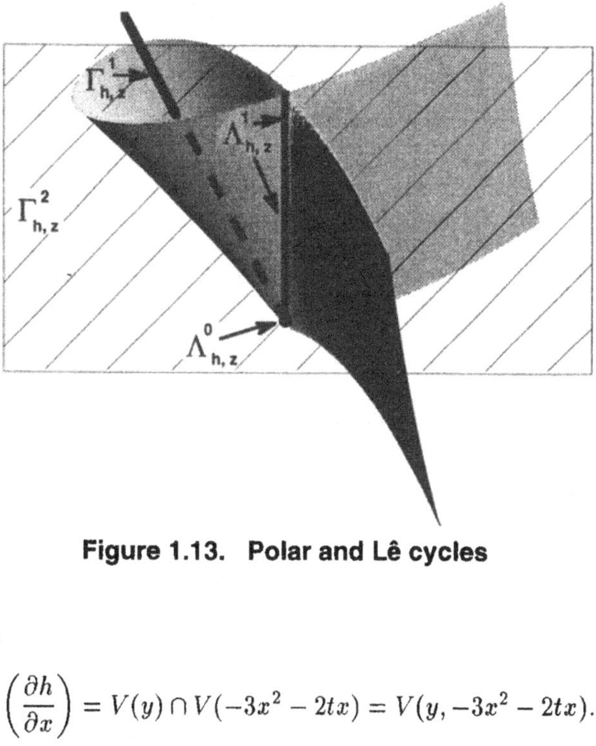

Figure 1.13. Polar and Lê cycles

Next, we have

$$\Gamma^2_h \cap V\left(\frac{\partial h}{\partial x}\right) = V(y) \cap V(-3x^2 - 2tx) = V(y, -3x^2 - 2tx).$$

Applying 1.4 and 1.2.iii, we find

$$\Gamma^1_h = \left(\Gamma^2_h \cap V\left(\frac{\partial h}{\partial x}\right)\right) \Big/ \Sigma h = V(y, -3x^2 - 2tx)/V(x, y) = V(y, -3x - 2t).$$

From the definition of the Lê cycles (1.11), we obtain

$$\Lambda^1_h = \left[V(y, -3x^2 - 2tx)\right] - \left[V(y, -3x - 2t)\right] =$$

$$\Big([V(y,x)] + [V(y,-3x-2t)]\Big) - [V(y,-3x-2t)] = [V(y,x)] .$$

Thus, Λ^1_h has as its underlying set the t-axis, and this axis occurs with multiplicity 1.

Now we find

$$\Lambda^0_h = \left[\Gamma^1_h \cap V\left(\frac{\partial h}{\partial t}\right)\right] = [V(y,-3x-2t)\cap V(-x^2)] = 2\,[V(t,x,y)] = 2[\mathbf{0}].$$

Finally, we calculate the Lê numbers: $\lambda^1_h(\mathbf{0}) = (V(y,x)\cdot V(t))_{\mathbf{0}} = 1$ and clearly $\lambda^0_h(\mathbf{0}) = 2$.

We will now prove some general results on Lê cycles and Lê numbers. In particular, we want to obtain Proposition 1.18, which will tell us that, when calculating Lê cycles, we can work solely on the cycle level, instead of worrying about the structure of the polar varieties as schemes.

Proposition 1.14. *The Lê cycles are all non-negative and are contained in the critical set of h. Every component of $\left|\Lambda^k_{h,\mathbf{z}}\right|$ has dimension at least k. If $s = \dim_{\mathbf{p}}\Sigma h$ then, for all k with $s < k < n+1$, $\mathbf{p}$ is not contained in $\left|\Lambda^k_{h,\mathbf{z}}\right|$, i.e. $\left|\Lambda^k_{h,\mathbf{z}}\right|$ is empty at $\mathbf{p}$; thus, for $s < k < n+1$, $\lambda^k_{h,\mathbf{z}}(\mathbf{p})$ is defined and equal to 0.*

Proof. The first statement follows from 1.4. The second statement follows from the definition of the Lê cycles and the fact that every component of $\Gamma^{k+1}_{h,\mathbf{z}}$ has dimension at least $k+1$. The third statement follows from the first two. □

Due to the proposition above, we usually only consider $\lambda^0, \dots, \lambda^s$.

Proposition 1.15. *As sets, for all k, $\Gamma^{k+1}_{h,\mathbf{z}} \cap \Sigma h = \bigcup_{i\leqslant k} \Lambda^i_{h,\mathbf{z}}$. In particular, letting $k = s := \dim_{\mathbf{p}}\Sigma h$, as germs of sets at $\mathbf{p}$, $\Sigma h = \bigcup_{i\leqslant s} \Lambda^i_{h,\mathbf{z}}$.*

Proof. The proof is by induction using the formula

$$\Gamma^{k+1}_{h,\mathbf{z}} \cap \Sigma h = \Gamma^{k+1}_{h,\mathbf{z}} \cap V\left(\frac{\partial h}{\partial z_k}\right) \cap \Sigma h =$$

$$(\Gamma^k_{h,\mathbf{z}} \cup \Lambda^k_{h,\mathbf{z}}) \cap \Sigma h = (\Gamma^k_{h,\mathbf{z}} \cap \Sigma h) \cup \Lambda^k_{h,\mathbf{z}}. \; \square$$

Recall from Remark 1.10 that, if $\gamma^1_{h,\mathbf{z}}(\mathbf{p})$ is defined and $\mathbf{p} \in \Sigma h$, then

$$\Sigma(h_{|_{V(z_0-p_0)}}) = V(z_0 - p_0) \cap \Sigma h$$

at $\mathbf{p}$. This is especially useful for inductive proofs when combined with the easy:

Proposition 1.16. *If $s := \dim_{\mathbf{p}} \Sigma h \geqslant 1$, $\Lambda^i_{h,\mathbf{z}}$ has correct dimension at $\mathbf{p}$ for all $i \leqslant s-1$ (by this, we mean to allow that $\mathbf{p}$ may not be contained in some of the $\Lambda^j_{h,\mathbf{z}}$'s), and $\lambda^s_{h,\mathbf{z}}(\mathbf{p})$ is defined, then $\dim_{\mathbf{p}}(\Sigma h \cap V(z_0 - p_0)) = s - 1$.*

Proof. As we are assuming that $\Lambda^i_{h,\mathbf{z}}$ has correct dimension at $\mathbf{p}$ for all $i \leqslant s-1$, it follows from 1.15 that we have only to show that the hyperplane slice $V(z_0 - p_0)$ actually reduces the dimension of $\Lambda^s_{h,\mathbf{z}}$. But, this must be the case, since

$$\Lambda^s_{h,\mathbf{z}} \cap V(z_0 - p_0, \ldots, z_{s-1} - p_{s-1})$$

is zero-dimensional at $\mathbf{p}$. □

We need the following lemma, which is an immediate consequence of 7.1.b of [**Fu**].

Lemma 1.17. *If a hypersurface $V(h)$ contains no components, embedded or isolated, of a scheme $V(\alpha)$ through a point $\mathbf{p}$, then the cycles $[V(\alpha)] \cdot [V(h)]$ and $[V(\alpha) \cap V(h)]$ are equal at $\mathbf{p}$.*

Proposition 1.18. *If, for all j with $0 \leqslant j \leqslant k$, $\Lambda^j_{h,\mathbf{z}}$ is purely j-dimensional at $\mathbf{p}$, then $\Gamma^{k+1}_{h,\mathbf{z}} \cap V\left(\frac{\partial h}{\partial z_k}\right)$ is purely k-dimensional at $\mathbf{p}$, and so $\Gamma^{k+1}_{h,\mathbf{z}}$ is purely $(k+1)$-dimensional at $\mathbf{p}$ and the cycles*

$$\left[\Gamma^{k+1}_{h,\mathbf{z}} \cap V\left(\frac{\partial h}{\partial z_k}\right)\right] \quad \text{and} \quad \left[\Gamma^{k+1}_{h,\mathbf{z}}\right] \cdot \left[V\left(\frac{\partial h}{\partial z_k}\right)\right]$$

are equal at $\mathbf{p}$.

Moreover, in this case, every k-dimensional (isolated) component of the critical locus of h through $\mathbf{p}$ is contained in $\left|\Lambda^k_{h,\mathbf{z}}\right|$.

Proof. We prove the first claim by an easy induction on k. For $k = 0$, the statement is trivially seen to be true as $\left[\Lambda^0_{h,\mathbf{z}}\right] = \left[\Gamma^1_{h,\mathbf{z}} \cap V\left(\frac{\partial h}{\partial z_0}\right)\right]$. Assume the hypothesis then for $k-1$, and suppose that $\Lambda^j_{h,\mathbf{z}}$ is purely j-dimensional at $\mathbf{p}$ for all j with $0 \leqslant j \leqslant k$. By our inductive hypothesis, $\Gamma^k_{h,\mathbf{z}}$ is purely k-dimensional at $\mathbf{p}$. Since, as sets, $\Gamma^{k+1}_{h,\mathbf{z}} \cap V\left(\frac{\partial h}{\partial z_k}\right) = \Gamma^k_{h,\mathbf{z}} \cup \Lambda^k_{h,\mathbf{z}}$ and we are assuming $\Lambda^k_{h,\mathbf{z}}$

is purely k-dimensional at $\mathbf{p}$, it is immediate that $\Gamma^{k+1}_{h,\mathbf{z}} \cap V\left(\frac{\partial h}{\partial z_k}\right)$ is purely k-dimensional at $\mathbf{p}$.

Now, assume that $\Gamma^{k+1}_{h,\mathbf{z}} \cap V\left(\frac{\partial h}{\partial z_k}\right)$ is purely k-dimensional at $\mathbf{p}$. Then it is immediate that $\Gamma^{k+1}_{h,\mathbf{z}}$ is purely $(k+1)$-dimensional at $\mathbf{p}$ and that $V\left(\frac{\partial h}{\partial z_k}\right)$ contains no (isolated) components of $\Gamma^{k+1}_{h,\mathbf{z}}$ through $\mathbf{p}$. Now, by Proposition 1.7, since $\Gamma^{k+1}_{h,\mathbf{z}}$ is purely $(k+1)$-dimensional at $\mathbf{p}$, $\Gamma^{k+1}_{h,\mathbf{z}}$ has no embedded components through $\mathbf{p}$. Thus, $V\left(\frac{\partial h}{\partial z_k}\right)$ contains no components – isolated or embedded – of $\Gamma^{k+1}_{h,\mathbf{z}}$ through $\mathbf{p}$, and hence we may use Lemma 1.17 to conclude that the cycles $\left[\Gamma^{k+1}_{h,\mathbf{z}} \cap V\left(\frac{\partial h}{\partial z_k}\right)\right]$ and $\left[\Gamma^{k+1}_{h,\mathbf{z}}\right] \cdot \left[V\left(\frac{\partial h}{\partial z_k}\right)\right]$ are equal at $\mathbf{p}$.

Finally, let C be a k-dimensional component of $|\Sigma h|$. Then, $C \subseteq C'$, where C' is a component of $\left|V\left(\frac{\partial h}{\partial z_{k+1}}, \ldots, \frac{\partial h}{\partial z_n}\right)\right|$. Now, as C is a component of $|\Sigma h|$ and $\dim C' \geqslant k+1$, we conclude that $C' \not\subseteq |\Sigma h|$, i.e. $C' \subseteq \left|\Gamma^{k+1}_{h,\mathbf{z}}\right|$. Therefore, as sets, we have

$$C = C \cap V\left(\frac{\partial h}{\partial z_k}\right) \subseteq \Gamma^{k+1}_{h,\mathbf{z}} \cap V\left(\frac{\partial h}{\partial z_k}\right) = \Gamma^{k}_{h,\mathbf{z}} \cup \Lambda^{k}_{h,\mathbf{z}}.$$

The conclusion now follows from the first part of the proposition. □

In practice, we use the first part of 1.18 to calculate the Lê cycles as follows:

Assume for the moment that all the Lê cycles have the correct dimension, and let s denote the dimension of the the critical locus of h. Then,

$$\left[\Gamma^{s+1}_{h,\mathbf{z}}\right] = \left[V\left(\frac{\partial h}{\partial z_{s+1}}, \ldots, \frac{\partial h}{\partial z_n}\right)\right].$$

$$\left[\Gamma^{s+1}_{h,\mathbf{z}}\right] \cdot \left[V\left(\frac{\partial h}{\partial z_s}\right)\right] = \left[\Gamma^{s}_{h,\mathbf{z}}\right] + \left[\Lambda^{s}_{h,\mathbf{z}}\right].$$

$$\left[\Gamma^{s}_{h,\mathbf{z}}\right] \cdot \left[V\left(\frac{\partial h}{\partial z_{s-1}}\right)\right] = \left[\Gamma^{s-1}_{h,\mathbf{z}}\right] + \left[\Lambda^{s-1}_{h,\mathbf{z}}\right].$$

$$\vdots$$

$$\left[\Gamma^{2}_{h,\mathbf{z}}\right] \cdot \left[V\left(\frac{\partial h}{\partial z_1}\right)\right] = \left[\Gamma^{1}_{h,\mathbf{z}}\right] + \left[\Lambda^{1}_{h,\mathbf{z}}\right].$$

$$\left[\Gamma^{1}_{h,\mathbf{z}}\right] \cdot \left[V\left(\frac{\partial h}{\partial z_0}\right)\right] = \left[\Lambda^{0}_{h,\mathbf{z}}\right].$$

In each line above, one obtains $\left[\Gamma^k_{h,\mathbf{z}}\right]$ from the calculation in the previous line.

Now, to write the above equalities, we have used that the Lê cycles have correct dimension – but, in any case, the equalities are true for **sets** (using intersection and union of sets, of course). And so, after doing the above calculations, one verifies that the cycles that we have written above as the Lê cycles do, in fact, have the correct dimension and thus the equalities are correct. On the other hand, if the cycles that we have written above as the Lê cycles do **not** have the correct dimension, then the equalities above may be false.

Remark 1.19. As we will see in Example 2.1, in the case of an isolated singularity, $\lambda^0_{h,\mathbf{z}}$ is nothing other than the Milnor number.

In the general case, it is tempting to think of $\lambda^0_{h,\mathbf{z}}(\mathbf{p})$ as the local (generic) degree of the Jacobian map of h at $\mathbf{p}$, i.e. the number of points in

$$\overset{\circ}{B}_\epsilon \cap V\left(\frac{\partial h}{\partial z_0} - a_0, \ldots, \frac{\partial h}{\partial z_n} - a_n\right),$$

where $\overset{\circ}{B}_\epsilon$ is a small open ball centered at $\mathbf{p}$ and $\mathbf{a}$ is a generic point with length that is small compared to ϵ; unfortunately, there is no such local degree.

Consider the example $h = z_2^2 + (z_0 - z_1^2)^2$ and let $\mathbf{p}$ be the origin.

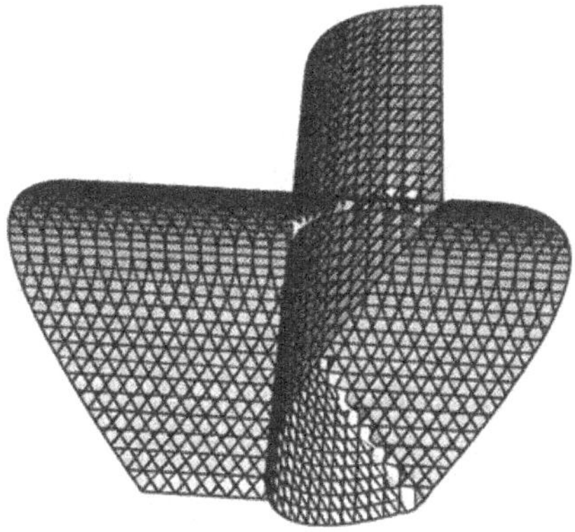

Figure 1.20. The hypersurface defined by h

Then,

$$\overset{\circ}{B}_\epsilon \cap V\left(\frac{\partial h}{\partial z_0} - a_0,\ \frac{\partial h}{\partial z_1} - a_1,\ \frac{\partial h}{\partial z_2} - a_2\right) =$$

$$\overset{\circ}{B}_\epsilon \cap V(2(z_0 - z_1^2) - a_0,\ 2(z_0 - z_1^2)(-2z_1) - a_1,\ 2z_2 - a_2).$$

The solutions to these equations are

$$z_0 = \frac{a_0}{2} + \frac{a_1^2}{4a_0^2}, \qquad z_1 = -\frac{a_1}{2a_0}, \qquad z_2 = \frac{a_2}{2}.$$

The number of solutions of these equations inside any small ball does not just depend on picking small, generic a_0, a_1, and a_2, but also depends on the relative sizes of a_0 and a_1. If a_1 is small relative to a_0, then there will be one solution inside the ball; if a_0 is small relative to a_1, then there will be no solutions inside the ball.

Do either of these numbers actually agree with $\lambda^0_{h,\mathbf{z}}(\mathbf{0})$? Yes; with these coordinates, $\lambda^0_{h,\mathbf{z}}(\mathbf{0}) = 1$. This can be seen from the above calculations together with the discussion below, which shows how "close" $\lambda^0_{h,\mathbf{z}}$ is to being the generic degree of the Jacobian map of h.

We claim that, if $\dim_{\mathbf{p}}\Gamma^1_{h,\mathbf{z}} = 1$, then $\lambda^0_{h,\mathbf{z}}(\mathbf{p})$ exists and equals the number of points in

$$\overset{\circ}{B}_\epsilon \cap V\left(\frac{\partial h}{\partial z_0} - a_0, \ldots, \frac{\partial h}{\partial z_n} - a_n\right),$$

where $\overset{\circ}{B}_\epsilon$ is a small open ball centered at $\mathbf{p}$, $a_0 \neq 0$ is small compared to ϵ, and $a_1, \ldots, a_n$ are generic, with length that is small compared to that of a_0.

To see this, note that this number of points equals the sum of the intersection numbers given by

$$\sum_{\mathbf{q}} \left(V\left(\frac{\partial h}{\partial z_1}, \ldots, \frac{\partial h}{\partial z_n}\right) \cdot V\left(\frac{\partial h}{\partial z_0} - a_0\right)\right)_{\mathbf{q}}$$

where the sum is over all $\mathbf{q}$ in

$$\overset{\circ}{B}_\epsilon \cap V\left(\frac{\partial h}{\partial z_0} - a_0, \frac{\partial h}{\partial z_1}, \ldots, \frac{\partial h}{\partial z_n}\right).$$

But, for $a_0 \neq 0$, these points, $\mathbf{q}$, do not occur on the critical locus of h, and so this sum equals

$$\sum_{\mathbf{q}} \left(\Gamma^1_{h,\mathbf{z}} \cdot V\left(\frac{\partial h}{\partial z_0} - a_0\right)\right)_{\mathbf{q}}.$$

This last sum is none other than

$$\lambda^0_{h,\mathbf{z}}(\mathbf{p}) = \left(\Gamma^1_{h,\mathbf{z}} \cdot V\left(\frac{\partial h}{\partial z_0}\right)\right)_{\mathbf{p}}.$$

It is also possible to give a more intuitive characterization of $\lambda^s_{h,\mathbf{z}}(\mathbf{p})$ where $s = \dim_{\mathbf{p}}\Sigma h$. Assuming that $\lambda^s_{h,\mathbf{z}}(\mathbf{p})$ exists, by moving to a generic point, it is trivial to show that

$$\lambda^s_{h,\mathbf{z}}(\mathbf{p}) = \sum_\nu n_\nu \overset{\circ}{\mu}_\nu,$$

where ν runs over all s-dimensional components of Σh at $\mathbf{p}$, n_ν is the local degree of the map $(z_0, \dots, z_{s-1})$ restricted to ν at $\mathbf{p}$, and $\overset{\circ}{\mu}_\nu$ denotes the generic transverse Milnor number of h along the component ν in a neighborhood of $\mathbf{p}$. In particular, if the coordinate system is generic enough so that n_ν is actually the multiplicity of ν at $\mathbf{p}$ for all ν, then $\lambda^s_{h,\mathbf{z}}(\mathbf{p})$ is merely the multiplicity of the Jacobian scheme of h (the scheme defined by the vanishing of the Jacobian ideal) at $\mathbf{p}$.

The next proposition tells us how the Lê numbers behave under the taking of hyperplane sections – a fundamental result. It is a combination of Propositions 2.5, 2.6, and 2.9 of [**Mas2**]; however, we have corrected an indexing error.

Proposition 1.21. *Suppose* $\Sigma h \cap V(z_0 - p_0) = \Sigma(h_{|_{V(z_0-p_0)}})$, *and use the coordinates* $\tilde{\mathbf{z}} = (z_1, \dots, z_n)$ *for* $V(z_0 - p_0)$. *Let* $k \geqslant 1$ *and suppose that* $\gamma^k_{h,\mathbf{z}}(\mathbf{p})$ *and* $\lambda^k_{h,\mathbf{z}}(\mathbf{p})$ *are defined.*

Then, $\gamma^{k-1}_{h_{|_{V(z_0-p_0)}},\tilde{\mathbf{z}}}(\mathbf{p})$ *and* $\lambda^{k-1}_{h_{|_{V(z_0-p_0)}},\tilde{\mathbf{z}}}(\mathbf{p})$ *are defined,*

$$\Gamma^k_{h_{|_{V(z_0-p_0)}},\tilde{\mathbf{z}}} = \Gamma^{k+1}_{h,\mathbf{z}} \cdot V(z_0 - p_0),$$

$$\Gamma^{k-1}_{h_{|_{V(z_0-p_0)}},\tilde{\mathbf{z}}} + \Lambda^{k-1}_{h_{|_{V(z_0-p_0)}},\tilde{\mathbf{z}}} = \left(\Gamma^k_{h,\mathbf{z}} + \Lambda^k_{h,\mathbf{z}}\right) \cdot V(z_0 - p_0),$$

and

$$\gamma^{k-1}_{h_{|_{V(z_0-p_0)}},\tilde{\mathbf{z}}}(\mathbf{p}) + \lambda^{k-1}_{h_{|_{V(z_0-p_0)}},\tilde{\mathbf{z}}}(\mathbf{p}) = \gamma^k_{h,\mathbf{z}}(\mathbf{p}) + \lambda^k_{h,\mathbf{z}}(\mathbf{p}).$$

In the special case when $k = 1$, *it follows that if* $\gamma^1_{h,\mathbf{z}}(\mathbf{p})$ *and* $\lambda^1_{h,\mathbf{z}}(\mathbf{p})$ *are defined, then so is* $\lambda^0_{h_{|_{V(z_0-p_0)}},\tilde{\mathbf{z}}}(\mathbf{p})$, *and*

$$\lambda^0_{h_{|_{V(z_0-p_0)}},\tilde{\mathbf{z}}}(\mathbf{p}) = \gamma^1_{h,\mathbf{z}}(\mathbf{p}) + \lambda^1_{h,\mathbf{z}}(\mathbf{p}).$$

Moreover, we conclude that if $k \geqslant 1$ *and* $\gamma^k_{h,\mathbf{z}}(\mathbf{p})$, $\lambda^k_{h,\mathbf{z}}(\mathbf{p})$, $\gamma^{k+1}_{h,\mathbf{z}}(\mathbf{p})$, *and* $\lambda^{k+1}_{h,\mathbf{z}}(\mathbf{p})$ *are defined, then so are*

$$\gamma^{k-1}_{h_{|_{V(z_0-p_0)}},\tilde{\mathbf{z}}}(\mathbf{p}),\ \lambda^{k-1}_{h_{|_{V(z_0-p_0)}},\tilde{\mathbf{z}}}(\mathbf{p}),\ \gamma^{k}_{h_{|_{V(z_0-p_0)}},\tilde{\mathbf{z}}}(\mathbf{p}),\quad \text{and}\ \lambda^{k}_{h_{|_{V(z_0-p_0)}},\tilde{\mathbf{z}}}(\mathbf{p}),$$

and

$$\Gamma^k_{h_{|_{V(z_0-p_0)}},\tilde{\mathbf{z}}} = \Gamma^{k+1}_{h,\mathbf{z}} \cdot V(z_0 - p_0),$$

$$\Lambda^k_{h_{|_{V(z_0-p_0)}},\tilde{\mathbf{z}}} = \Lambda^{k+1}_{h,\mathbf{z}} \cdot V(z_0 - p_0),$$

and so

$$\gamma^k_{h_{|V(z_0-p_0)}, \tilde{\mathbf{z}}}(\mathbf{p}) = \gamma^{k+1}_{h,\mathbf{z}}(\mathbf{p}),$$

and

$$\lambda^k_{h_{|V(z_0-p_0)}, \tilde{\mathbf{z}}}(\mathbf{p}) = \lambda^{k+1}_{h,\mathbf{z}}(\mathbf{p}).$$

Proof. Clearly, it suffices to prove the assertions for $\mathbf{p} = \mathbf{0}$. The assumption that $\gamma^k_{h,\mathbf{z}}(\mathbf{0})$ and $\lambda^k_{h,\mathbf{z}}(\mathbf{0})$ are defined is equivalent to

$$\dim_{\mathbf{0}} \Gamma^{k+1}_{h,\mathbf{z}} \cap V\left(\frac{\partial h}{\partial z_k}\right) \cap V(z_0, \dots, z_{k-1}) \leqslant 0.$$

Hence, $\Gamma^{k+1}_{h,\mathbf{z}}$ is purely $(k+1)$-dimensional at the origin and thus has no embedded components (Proposition 1.7). Also, $\Gamma^{k+1}_{h,\mathbf{z}} \cap V\left(\frac{\partial h}{\partial z_k}\right)$ is purely k-dimensional at the origin and so, by Lemma 1.17, we have an equality of cycles

$$\left[\Gamma^{k+1}_{h,\mathbf{z}} \cap V\left(\frac{\partial h}{\partial z_k}\right)\right] = \Gamma^{k+1}_{h,\mathbf{z}} \cdot V\left(\frac{\partial h}{\partial z_k}\right) = \Gamma^k_{h,\mathbf{z}} + \Lambda^k_{h,\mathbf{z}}.$$

In addition, we see that $\Gamma^{k+1}_{h,\mathbf{z}} \cap V\left(\frac{\partial h}{\partial z_k}\right) \cap V(z_0)$ is purely $(k-1)$-dimensional at the origin; we easily conclude that

$$\dim_{\mathbf{0}} \Gamma^{k+1}_{h,\mathbf{z}} \cap \Sigma h \cap V(z_0) \leqslant k-1.$$

Now, let us consider the cycle $\Gamma^k_{h_{|V(z_0)}, \tilde{\mathbf{z}}}$. By definition,

$$\Gamma^k_{h_{|V(z_0)}, \tilde{\mathbf{z}}} = V\left(z_0, \frac{\partial h}{\partial z_{k+1}}, \dots, \frac{\partial h}{\partial z_n}\right) \Big/ \Sigma(h_{|V(z_0)}).$$

Using Lemma 1.2.ii and our hypothesis that $\Sigma h \cap V(z_0) = \Sigma(h_{|V(z_0)})$, the equality above gives us

$$\Gamma^k_{h_{|V(z_0)}, \tilde{\mathbf{z}}} = \left(V(z_0) \cap \Gamma^{k+1}_{h,\mathbf{z}}\right) \Big/ (\Sigma h \cap V(z_0)) = \left(V(z_0) \cap \Gamma^{k+1}_{h,\mathbf{z}}\right) \Big/ \Sigma h.$$

But, $V(z_0) \cap \Gamma^{k+1}_{h,\mathbf{z}}$ is purely k-dimensional at the origin and, as we saw earlier, $\dim_{\mathbf{0}} \Gamma^{k+1}_{h,\mathbf{z}} \cap \Sigma h \cap V(z_0) \leqslant k-1$; therefore, Σh contains no isolated components of $V(z_0) \cap \Gamma^{k+1}_{h,\mathbf{z}}$ and so, as cycles,

$$\Gamma^k_{h_{|V(z_0)}, \tilde{\mathbf{z}}} = \Gamma^{k+1}_{h,\mathbf{z}} \cap V(z_0) = \Gamma^{k+1}_{h,\mathbf{z}} \cdot V(z_0).$$

We find

$$\Gamma^{k-1}_{h_{|V(z_0)}, \tilde{\mathbf{z}}} + \Lambda^{k-1}_{h_{|V(z_0)}, \tilde{\mathbf{z}}} = \Gamma^k_{h_{|V(z_0)}, \tilde{\mathbf{z}}} \cdot V\left(\frac{\partial h}{\partial z_k}\right) =$$

$$\Gamma^{k+1}_{h,\mathbf{z}} \cdot V(z_0) \cdot V\left(\frac{\partial h}{\partial z_k}\right) = \left(\Gamma^k_{h,\mathbf{z}} + \Lambda^k_{h,\mathbf{z}}\right) \cdot V(z_0).$$

That $\gamma^{k-1}_{h_{|V(z_0)},\tilde{\mathbf{z}}}(\mathbf{0})$ and $\lambda^{k-1}_{h_{|V(z_0)},\tilde{\mathbf{z}}}(\mathbf{0})$ are defined and that

$$\gamma^{k-1}_{h_{|V(z_0)},\tilde{\mathbf{z}}}(\mathbf{0}) + \lambda^{k-1}_{h_{|V(z_0)},\tilde{\mathbf{z}}}(\mathbf{0}) = \gamma^k_{h,\mathbf{z}}(\mathbf{0}) + \lambda^k_{h,\mathbf{z}}(\mathbf{0})$$

follows by intersecting the cycle $V(z_1, \dots, z_n)$ with each side of the above equality of cycles.

The remaining equalities follow easily – we leave them as an exercise. □

Corollary 1.22. *Let $k \geqslant 0$. Suppose*

$$\Sigma h \cap V(z_0 - p_0, \dots, z_k - p_k) = \Sigma(h_{|V(z_0 - p_0, \dots, z_k - p_k)}),$$

and that $\gamma^i_{h,\mathbf{z}}(\mathbf{p})$ and $\lambda^i_{h,\mathbf{z}}(\mathbf{p})$ are defined for all $i \leqslant k$. Then, $\gamma^{k+1}_{h,\mathbf{z}}(\mathbf{p})$ is defined.

Proof. Again, we prove this for $\mathbf{p} = \mathbf{0}$.

Case 1: If $\mathbf{0} \notin \Sigma h$, then near $\mathbf{0}$,

$$\Gamma^{k+1}_{h,\mathbf{z}} = V\left(\frac{\partial h}{\partial z_{k+1}}, \dots, \frac{\partial h}{\partial z_n}\right)$$

and so

$$\Gamma^{k+1}_{h,\mathbf{z}} \cap V(z_0, \dots z_k) = \Sigma(h_{|V(z_0, \dots, z_k)}) = \Sigma h \cap V(z_0, \dots, z_k) = \emptyset.$$

Hence, $\gamma^{k+1}_{h,\mathbf{z}}(\mathbf{0})$ is defined and equal to zero.

Case 2: $\mathbf{0} \in \Sigma h$. The proof is by induction on k.

For $k = 0$, the claim is that if $\mathbf{0} \in \Sigma h$, $\lambda^0_{h,\mathbf{z}}(\mathbf{0})$ is defined, and $\Sigma h \cap V(z_0) = \Sigma(h_{|V(z_0)})$, then $\dim_{\mathbf{0}} \Gamma^1_{h,\mathbf{z}} \cap V(z_0) \leqslant 0$. As $\mathbf{0} \in \Sigma h$ and $\lambda^0_{h,\mathbf{z}}(\mathbf{0})$ is defined, we must have that $\dim_{\mathbf{0}} \Gamma^1_{h,\mathbf{z}} \leqslant 1$. So, if $\dim_{\mathbf{0}} \Gamma^1_{h,\mathbf{z}} \cap V(z_0) \geqslant 1$, then $V(z_0)$ must contain a component of $\Gamma^1_{h,\mathbf{z}}$ through the origin. But, since $\Sigma h \cap V(z_0) = \Sigma(h_{|V(z_0)})$,

$$\Gamma^1_{h,\mathbf{z}} \cap V(z_0) \subseteq V\left(z_0, \frac{\partial h}{\partial z_1}, \dots, \frac{\partial h}{\partial z_n}\right) = V\left(z_0, \frac{\partial h}{\partial z_0}, \dots, \frac{\partial h}{\partial z_n}\right).$$

Hence, any component of $\Gamma^1_{h,\mathbf{z}}$ contained in $V(z_0)$ must also be contained in Σh; this contradicts the definition of $\Gamma^1_{h,\mathbf{z}}$.

Suppose now that the corollary is true up to $k - 1$, where $k \geqslant 1$. Suppose $\mathbf{0} \in \Sigma h$, $\Sigma h \cap V(z_0, \dots, z_k) = \Sigma(h_{|V(z_0, \dots, z_k)})$, and that $\gamma^i_{h,\mathbf{z}}(\mathbf{0})$ and $\lambda^i_{h,\mathbf{z}}(\mathbf{0})$ are defined for all $i \leqslant k$. As $\mathbf{0} \in \Sigma h$ and $\gamma^i_{h,\mathbf{z}}(\mathbf{0})$ is defined for all $i \leqslant k$, Remark 1.10 implies that $\Sigma h \cap V(z_0, \dots, z_i) = \Sigma(h_{|_{V(z_0,\dots,z_i)}})$ for all $i \leqslant k-1$.

In particular, as $k \geqslant 1$, $\Sigma h \cap V(z_0) = \Sigma(h_{|_{V(z_0)}})$. Thus, we may apply Proposition 1.21 to conclude that $\gamma^i_{h_{|_{V(z_0)}},\tilde{\mathbf{z}}}(\mathbf{0})$ and $\lambda^i_{h_{|_{V(z_0)}},\tilde{\mathbf{z}}}(\mathbf{0})$ are defined for all $i \leqslant k-1$ and, as sets,

$$\Gamma^k_{h_{|_{V(z_0)}},\tilde{\mathbf{z}}} = \Gamma^{k+1}_{h,\mathbf{z}} \cap V(z_0).$$

Since $\Sigma(h_{|_{V(z_0)}}) \cap V(z_1, \dots, z_k) = \Sigma(h_{|_{V(z_0,\dots,z_k)}})$, we are in a position to apply our inductive hypothesis to $h_{|_{V(z_0)}}$.

We conclude that $\gamma^k_{h_{|_{V(z_0)}},\tilde{\mathbf{z}}}(\mathbf{0})$ is defined, i.e.

$$\dim_{\mathbf{0}} \Gamma^k_{h_{|_{V(z_0)}},\tilde{\mathbf{z}}} \cap V(z_1, \dots, z_k) \leqslant 0.$$

As $\Gamma^k_{h_{|_{V(z_0)}},\tilde{\mathbf{z}}} = \Gamma^{k+1}_{h,\mathbf{z}} \cap V(z_0)$, the proof is finished. □

We shall need the following relation between three intersection numbers.

Proposition 1.23. *Let* $\mathbf{p} \in \Sigma h$. *Then,* $\lambda^0_{h,\mathbf{z}}(\mathbf{p})$ *is defined if and only if* $\dim_{\mathbf{p}} \Gamma^1_{h,\mathbf{z}} \leqslant 1$.

Moreover, if $\gamma^1_{h,\mathbf{z}}(\mathbf{p})$ *is defined, then* $\lambda^0_{h,\mathbf{z}}(\mathbf{p})$ *is defined, the dimension of* $\Gamma^1_{h,\mathbf{z}} \cap V(h - h(\mathbf{p}))$ *at* $\mathbf{p}$ *is at most zero, and*

$$\left(\Gamma^1_{h,\mathbf{z}} \cdot V(h - h(\mathbf{p}))\right)_{\mathbf{p}} = \lambda^0_{h,\mathbf{z}}(\mathbf{p}) + \gamma^1_{h,\mathbf{z}}(\mathbf{p}).$$

Proof. $\Gamma^1_{h,\mathbf{z}}$ consists of those components of $V\left(\frac{\partial h}{\partial z_1}, \dots, \frac{\partial h}{\partial z_n}\right)$ which are not contained in $|\Sigma h|$. Thus, $V\left(\frac{\partial h}{\partial z_0}\right)$ contains no components of $\Gamma^1_{h,\mathbf{z}}$. Therefore, $\Gamma^1_{h,\mathbf{z}}$ is purely one-dimensional at $\mathbf{p}$ if and only if $\Gamma^1_{h,\mathbf{z}} \cap V\left(\frac{\partial h}{\partial z_0}\right)$ is purely zero-dimensional at $\mathbf{p}$, i.e. if and only if $\lambda^0_{h,\mathbf{z}}(\mathbf{p})$ is defined.

If $\gamma^1_{h,\mathbf{z}}(\mathbf{p})$ is defined, then $\Gamma^1_{h,\mathbf{z}}$ must be purely one-dimensional at $\mathbf{p}$ and so $\lambda^0_{h,\mathbf{z}}(\mathbf{p})$ is defined, by the above.

We now follow the proof in Proposition 1.3 of [**Lê1**] to show that an easy application of the chain rule yields $\dim_{\mathbf{p}} \Gamma^1_{h,\mathbf{z}} \cap V(h - h(\mathbf{p})) \leqslant 0$, and

$$\left(\Gamma^1_{h,\mathbf{z}} \cdot V(h - h(\mathbf{p}))\right)_{\mathbf{p}} = \lambda^0_{h,\mathbf{z}}(\mathbf{p}) + \gamma^1_{h,\mathbf{z}}(\mathbf{p}).$$

For convenience, we assume that $\mathbf{p} = \mathbf{0}$ and that $h(\mathbf{0}) = 0$.

Suppose $\Gamma^1_{h,\mathbf{z}} = \sum m_W[W]$ as cycles. We know that we can calculate the intersection number of a curve and a hypersurface by parameterizing the curve and looking at the multiplicity of the composition of the defining function of the hypersurface with the parameterization. So, for each component W, pick a local analytic parameterization $\alpha(t)$ of W such that $\alpha(0) = \mathbf{0}$. We must show two things: that $h(\alpha(t))$ is not identically zero, and that

$$\text{mult}_t h(\alpha(t)) = \text{mult}_t \left(\frac{\partial h}{\partial z_0}\right)_{|\alpha(t)} + \text{mult}_t z_0(\alpha(t)).$$

As we already know that the righthand side of the above equality is finite, we have only to prove that the equality holds in order to conclude that $h(\alpha(t))$ is not identically zero. But this is easy:

$$\text{mult}_t h(\alpha(t)) = 1 + \text{mult}_t \frac{d}{dt}\left\{h(\alpha(t))\right\} = 1 + \text{mult}_t \left\{\left(\frac{\partial h}{\partial z_0}\right)_{|\alpha(t)} \cdot \alpha_0'(t)\right\},$$

where the remaining terms that come from the chain rule are zero since $\alpha(t)$ parameterizes a component of the polar curve. Thus,

$$\text{mult}_t h(\alpha(t)) = 1 + \text{mult}_t \left(\frac{\partial h}{\partial z_0}\right)_{|\alpha(t)} + \text{mult}_t \alpha_0'(t)$$

$$= \text{mult}_t \left(\frac{\partial h}{\partial z_0}\right)_{|\alpha(t)} + \text{mult}_t \alpha_0(t) = \text{mult}_t \left(\frac{\partial h}{\partial z_0}\right)_{|\alpha(t)} + \text{mult}_t z_0(\alpha(t))$$

and we are finished. □

Of course, what we want to know is, for a generic choice of coordinates, $\mathbf{z}$, the polar numbers and the Lê numbers are actually defined. This is, in fact, the case. We pick the coordinates generically with respect to a certain type of stratification of the hypersurface defined by h.

Definition 1.24. If X is analytic space, an *analytic stratification* of X is a locally finite partition, $\{S_\alpha\}$, of X into analytic submanifolds – *the strata* – such the closure of each stratum is also analytic, and such that $\{S_\alpha\}$ satisfies the condition of the frontier, i.e. the closure of each stratum is a union of strata.

A *good stratification* for h at a point $\mathbf{p} \in V(h)$ is an analytic stratification, $\mathfrak{G}$, of the hypersurface $V(h)$ in a neighborhood, $\mathcal{U}$, of $\mathbf{p}$ such that the smooth part of $V(h)$ is a stratum and so that the stratification satisfies Thom's a_h condition with respect to $\mathcal{U} - V(h)$. That is, if $\mathbf{q}_i$ is a sequence of points in $\mathcal{U} - V(h)$ such that $\mathbf{q}_i \to \mathbf{q} \in S \in \mathfrak{G}$ and $T_{\mathbf{q}_i}V(h - h(\mathbf{q}_i))$ converges to some hyperplane $\mathcal{T}$, then $T_{\mathbf{q}}S \subseteq \mathcal{T}$.

Proposition 1.25 (Hamm and Lê [**H-L**]). *There exists a good stratification for all* $h : (\mathcal{U}, \mathbf{0}) \to (\mathbb{C}, 0)$ *at all* $\mathbf{p} \in V(h)$.

The notion defined below, that of *prepolar coordinates*, is crucial throughout the remainder of these notes. It provides a generic condition on linear choices of coordinates which implies that all the Lê numbers and polar numbers are defined. Moreover, prepolarity seems to be the right condition to obtain many topological results. The importance of this definition cannot be overstated.

Definition 1.26. Suppose that $\{S_\alpha\}$ is a good stratification for h in a neighborhood, $\mathcal{U}$, of the origin. Let $\mathbf{p} \in V(h)$. Then, a hyperplane, H, in $\mathbb{C}^{n+1}$ through $\mathbf{p}$ is a *prepolar slice* for h at $\mathbf{p}$ with respect to $\{S_\alpha\}$ provided that H transversely intersects all the strata of $\{S_\alpha\}$– except perhaps the stratum $\{\mathbf{p}\}$ itself – in a neighborhood of $\mathbf{p}$.

If H is a prepolar slice for h at $\mathbf{p}$ with respect to $\{S_\alpha\}$, then, as germs of sets at $\mathbf{p}$, $\Sigma(h_{|_H}) = (\Sigma h) \cap H$ and $\dim_{\mathbf{p}} \Sigma(h_{|_H}) = (\dim_{\mathbf{p}} \Sigma h) - 1$ provided $\dim_{\mathbf{p}} \Sigma h \geqslant 1$; moreover, $\{H \cap S_\alpha\}$ is a good stratification for $h_{|_H}$ at $\mathbf{p}$ (see [**H-L**]).

By 2.1.3 of [**H-L**], for a fixed good stratification for h, prepolar slices are generic.

We say simply that H is a *prepolar slice* for h at $\mathbf{p}$ provided that there exists a good stratification with respect to which H is a prepolar slice.

Let $(z_0, \dots, z_n)$ be a linear choice of coordinates for $\mathbb{C}^{n+1}$, let $\mathbf{p} \in V(h)$, and let $\{S_\alpha\}$ be a good stratification for h at $\mathbf{p}$.

For $0 \leqslant i \leqslant n$, $(z_0, \dots, z_i)$ is a *prepolar-tuple* for h at $\mathbf{p}$ with respect to $\{S_\alpha\}$ if and only if $V(z_0 - p_0)$ is a prepolar slice for h at $\mathbf{p}$ with respect to $\{S_\alpha\}$ and for all j such that $1 \leqslant j \leqslant i$, $V(z_j - p_j)$ is a prepolar slice for $h_{|_{V(z_0 - p_0, \dots, z_{j-1} - p_{j-1})}}$ at $\mathbf{p}$ with respect to $\{S_\alpha \cap V(z_0 - p_0, \dots, z_{j-1} - p_{j-1})\}$.

As prepolar slices are generic, so are prepolar-tuples.

Naturally, we say that $(z_0, \dots, z_i)$ is a *prepolar-tuple* for h at $\mathbf{p}$ provided that there exists a good stratification for h at $\mathbf{p}$ with respect to which $(z_0, \dots, z_i)$ is a prepolar-tuple.

Finally, we say that the coordinates $(z_0, \dots, z_n)$ are *prepolar* for h if and only if for all $\mathbf{p} \in V(h)$, if s denotes $\dim_{\mathbf{p}} \Sigma h$, then $(z_0, \dots, z_{s-1})$ is a prepolar-tuple for h at $\mathbf{p}$ (if $s = 0$ or $\mathbf{p} \notin \Sigma h$, we mean that there is no condition on the coordinates.)

Note that, as prepolar for h is a condition at **all** points in Σh, it is **not** immediate that such coordinates exist (we shall, however, prove this in Chapter 10.)

We will show that by choosing coordinates which are prepolar, one guarantees the existence of the Lê and polar numbers.

Lemma 1.27. *Let $L_k = (z_0, \ldots, z_k)$ be prepolar for h at the origin. Then, for all $\mathbf{p} \in V(h) \cap V(z_0, \ldots, z_{k-1})$ near $\mathbf{0}$, if $\mathcal{T}_\mathbf{p}$ is a hyperplane such that there exists a sequence $\mathbf{p}_i \notin \Sigma h$ with $\mathbf{p}_i \to \mathbf{p}$ and $T_{\mathbf{p}_i} V(h - h(\mathbf{p}_i)) \to \mathcal{T}_\mathbf{p}$, then $V(L_k) = T_\mathbf{p} V(L_k - L_k(\mathbf{p})) \not\subseteq \mathcal{T}_\mathbf{p}$.*

Proof. Fix a good stratification, $\mathfrak{G}$, with respect to which L_k is prepolar. Our proof is by induction on k. For $k = 0$, $V(z_0, \ldots, z_{k-1})$ equals the entire ambient space, $\mathbb{C}^{n+1}$, and the assertion follows immediately from the definition of a prepolar slice.

Suppose the lemma is true for k, but false for $k + 1$. Then, there must exist points, $\mathbf{p}$, arbitrarily close to the origin with the following properties: $\mathbf{p} \in V(h) \cap V(z_0, \ldots, z_k)$ and there exists a sequence $\mathbf{p}_i \notin \Sigma h$ with $\mathbf{p}_i \to \mathbf{p}$ and $T_{\mathbf{p}_i} V(h - h(\mathbf{p}_i))$ converging to some hyperplane $\mathcal{T}_\mathbf{p}$ with $V(z_0, \ldots, z_{k+1}) \subseteq \mathcal{T}_\mathbf{p}$. By taking a subsequence, we may assume that the $\mathbf{p}$'s are contained in a single good stratum, G. By our inductive hypothesis, we may also assume that $V(z_0, \ldots, z_k) \not\subseteq \mathcal{T}_\mathbf{p}$. Thus,

$$\dim \mathcal{T}_\mathbf{p} \cap V(z_0, \ldots, z_k) = n - k - 1 = \dim \mathcal{T}_\mathbf{p} \cap V(z_0, \ldots, z_{k+1}),$$

and so $\mathcal{T}_\mathbf{p} \cap V(z_0, \ldots, z_k) \subseteq V(z_{k+1})$. Therefore, $T_\mathbf{p}(G \cap V(z_0, \ldots, z_k)) = T_\mathbf{p}(G) \cap V(z_0, \ldots, z_k) \subseteq \mathcal{T}_\mathbf{p} \cap V(z_0, \ldots, z_k) \subseteq V(z_{k+1})$; a contradiction, as $V(z_{k+1})$ is a prepolar slice of $h_{|_{V(z_0, \ldots, z_k)}}$ at $\mathbf{0}$ with respect to the good stratification $\mathfrak{G} \cap V(z_0, \ldots, z_k)$. $\square$

Theorem 1.28. *Let $\mathbf{p} \in V(h)$ and let $s = \dim_\mathbf{p} \Sigma h$. Then, for a generic choice of coordinates, all of the Lê numbers and polar numbers of h at $\mathbf{p}$ in dimensions less than or equal to s are defined.*

More precisely, if $(z_0, \ldots, z_k)$ is prepolar for h at $\mathbf{p}$ then, for all i with $0 \leqslant i \leqslant k$, $\lambda^i_{h,\mathbf{z}}(\mathbf{p})$ and $\gamma^{i+1}_{h,\mathbf{z}}(\mathbf{p})$ exist. Moreover, if $(z_0, \ldots, z_{s-1})$ is prepolar for h at $\mathbf{p}$, then for all i with $0 \leqslant i \leqslant s$, $\lambda^i_{h,\mathbf{z}}(\mathbf{p})$ and $\gamma^i_{h,\mathbf{z}}(\mathbf{p})$ exist.

Proof. Note that, as prepolar slices transversely intersect the smooth stratum, Corollary 1.22 tells us that we only have to prove: if $(z_0, \ldots, z_k)$ is prepolar for h at $\mathbf{p}$ then, for all i with $0 \leqslant i \leqslant k$, $\lambda^i_{h,\mathbf{z}}(\mathbf{p})$ and $\gamma^i_{h,\mathbf{z}}(\mathbf{p})$ exist – for then the existence of $\gamma^{i+1}_{h,\mathbf{z}}(\mathbf{p})$ follows. Clearly it suffices to prove this when $\mathbf{p}$ is the origin.

If $(z_0, \ldots, z_k)$ is prepolar, then it follows from the definition that $(z_0, \ldots, z_i)$ is prepolar for all $i \leqslant k$. Hence, it suffices to prove that if $(z_0, \ldots, z_k)$ is prepolar, then $\lambda^k_{h,\mathbf{z}}(\mathbf{0})$ and $\gamma^k_{h,\mathbf{z}}(\mathbf{0})$ exist.

Let $L_{k-1} = (z_0, \ldots, z_{k-1})$ and $L_k = (z_0, \ldots, z_k)$. Let $\mathfrak{G}$ be a good stratification for h at the origin with respect to which L_k is prepolar. We wish to show that

$$\Gamma^{k+1}_{h,L_k} \cap V\left(\frac{\partial h}{\partial z_k}\right) \cap V(L_{k-1})$$

is 0-dimensional at the origin. We shall show that, in fact,

$$\Gamma^{k+1}_{h,L_k} \cap V(h) \cap V(L_{k-1})$$

is 0-dimensional at the origin. The desired conclusion then follows from the chain rule since any analytic curve in $\Gamma^{k+1}_{h,L_k} \cap V\left(\frac{\partial h}{\partial z_k}\right) \cap V(L_{k-1})$ which passes through the origin must be contained in $V(h)$.

Suppose that we had a sequence of $\mathbf{p}$ in $\Gamma^{k+1}_{h,L_k} \cap V(h) \cap V(L_{k-1})$ approaching $\mathbf{0}$. As each $\mathbf{p}$ is contained in Γ^{k+1}_{h,L_k}, for each $\mathbf{p}$ there exists a sequence $\mathbf{p}_i \to \mathbf{p}$ such that $\mathbf{p}_i \notin \Sigma h$ and $T_{\mathbf{p}_i}V(L_k - L_k(\mathbf{p}_i)) \subseteq T_{\mathbf{p}_i}V(h - h(\mathbf{p}_i))$. By taking a subsequence of the $\mathbf{p}_i$'s, we may also assume that the $T_{\mathbf{p}_i}V(h - h(\mathbf{p}_i))$ converge to some hyperplane, $\mathcal{T}_{\mathbf{p}}$. But, by construction, $V(L_k) = T_{\mathbf{p}}V(L_k - L_k(\mathbf{p})) \subseteq \mathcal{T}_{\mathbf{p}}$, which contradicts the lemma.

It remains for us to show that if $(z_0, \dots, z_{s-1})$ is prepolar for h at $\mathbf{0}$, then $\lambda^s_{h,\mathbf{z}}(\mathbf{0})$ exists; in other words, we get the last dimension without extra assumptions.

This is easy for $\Gamma^{s+1}_{h,L_s} = V\left(\frac{\partial h}{\partial z_{s+1}}, \dots, \frac{\partial h}{\partial z_n}\right)$, regardless of the choice of z_s, and so, as sets,

$$\Gamma^{s+1}_{h,L_s} \cap V\left(\frac{\partial h}{\partial z_s}\right) \cap V(z_0, \dots, z_{s-1}) = \Sigma(h_{|_{V(z_0,\dots,z_{s-1})}}).$$

Now, $\Sigma(h_{|_{V(z_0,\dots,z_{s-1})}})$ is easily seen to be 0-dimensional at the origin since each successive slice is prepolar. □

Remark 1.29. While it is true that prepolar coordinates occur generically and guarantee the existence of the Lê numbers, it is **not** true that all sets of prepolar coordinates yield the same Lê numbers.

If $\dim_{\mathbf{p}}\Sigma h = 1$, then $\{V(h) - \Sigma h, \Sigma h - \mathbf{p}, \mathbf{p}\}$ is a good stratification for h in a neighborhood of $\mathbf{p}$; hence, $V(z_0 - p_0)$ is a prepolar slice if and only if $\dim_{\mathbf{p}}\Sigma(h_{|_{V(z_0-p_0)}}) = 0$. This is the case if and only if $\gamma^1_{h,\mathbf{z}}(\mathbf{p})$ and $\lambda^1_{h,\mathbf{z}}(\mathbf{p})$ are defined.

Now, consider the example from Remark 1.19. The coordinates $\mathbf{z} = (z_0, z_1, z_2)$ are prepolar for $h = z_2^2 + (z_0 - z_1^2)^2$ at the origin, and $\lambda^0_{h,\mathbf{z}}(\mathbf{0}) = 1$ and $\lambda^1_{h,\mathbf{z}}(\mathbf{0}) = 2$. However, the coordinates $\mathbf{z}$ are really not very generic, as Σh is smooth at the origin, but $V(z_0)$ intersects Σh with multiplicity 2 at $\mathbf{0}$. The generic values of λ^0_h and λ^1_h (that is, the values with respect to generic coordinates) are 0 and 1, respectively.

Note that the alternating sum of the Lê numbers is the same for the non-generic and generic coordinates. As we shall see in Chapter 3, this is a general fact: as long as the coordinates are prepolar, the alternating sum of the Lê numbers is independent of the coordinates and is, in fact, equal to the reduced Euler characteristic of the Milnor fibre. We know of no algebraic way to prove this independence.

It is reasonable to ask why we do not strengthen our notion of prepolar in order to disallow examples such as the one above, where the Lê numbers do not have their generic values. The answer is that later (in Proposition 10.2) we shall show that, given h and a point $\mathbf{p} \in V(h)$, one may pick generic coordinates which are prepolar for h **at every point in a neighborhood of p**. This is the result that leads us into the more advanced results that are briefly described in Chapter 10. Example 2.4 shows that this result would be false if we were to strengthen the notion of prepolar to require the Lê numbers to obtain their generic values at each point in this open neighborhood of $\mathbf{p}$.

We conclude this chapter with four results which do not seem to be of fundamental importance, but which are fairly surprising.

Proposition 1.30. *Suppose that* $\dim_{\mathbf{p}}\Sigma h = 1$ *and* $V(z_0 - p_0)$ *is a prepolar slice for* h *at* $\mathbf{p}$. *If* $V(z_0 - p_0)$ *does not transversely intersect the set* $|\Sigma h|$ *at* $\mathbf{p}$ *(in particular, if* $|\Sigma h|$ *is not smooth at* $\mathbf{p}$*), then* $\lambda^0_{h,\mathbf{z}}(\mathbf{p}) \neq 0$.

Proof. Despite the different appearance of the statement, this is precisely what Lê proves in [**Lê8**]. □

Proposition 1.31. *Let* $k \geqslant 1$. *Suppose that* $\Lambda^0_{h,\mathbf{z}}, \ldots, \Lambda^{k-1}_{h,\mathbf{z}}$ *have correct dimension at* $\mathbf{p}$. *Suppose, for all pairs of distinct irreducible germs,* V *and* W, *of* Σh *through* $\mathbf{p}$, *that* $\dim_{\mathbf{p}}(V \cap W) \leqslant k-1$. *Finally, suppose that* $\lambda^k_{h,\mathbf{z}}(\mathbf{p}) = 0$. *Then,* $\lambda^j_{h,\mathbf{z}}(\mathbf{p}) = 0$ *for all* $j \leqslant k$.

Proof. One applies 2.3 of [**La**] to the case where the irreducible normal variety is $\mathbb{C}^{n+1}$ and the subvariety locally defined by $n-k$ equations is $V\left(\frac{\partial h}{\partial z_{k+1}}, \ldots, \frac{\partial h}{\partial z_n}\right)$, which equals $\Gamma^{k+1}_{h,\mathbf{z}} \cup \Sigma h$ as a set.

Let V be an irreducible component of Σh at $\mathbf{p}$ and let $\{W_i\}_i$ be the remaining irreducible components of Σh at $\mathbf{p}$.

Then, the lemma of Lazarsfeld says that if

$$\mathbf{p} \in \left(\Gamma^{k+1}_{h,\mathbf{z}} \cap V\right) \cup \left(\left(\bigcup_i W_i\right) \cap V\right),$$

then

$$\dim_{\mathbf{p}} \left(\Gamma^{k+1}_{h,\mathbf{z}} \cap V\right) \cup \left(\left(\bigcup_i W_i\right) \cap V\right) \geqslant k.$$

Now the proposition follows easily from 1.15. □

Proposition 1.32. *Let $s = \dim_{\mathbf{p}}\Sigma h$ and suppose that $\lambda^j_{h,\mathbf{z}}(\mathbf{p})$ and $\gamma^j_{h,\mathbf{z}}(\mathbf{p})$ exist for all $j \leqslant s$. Suppose that the critical locus of h at $\mathbf{p}$ is itself singular and denote the singular set of the critical locus by $\Sigma\Sigma h$. Then, every $(s-1)$-dimensional component of $\Sigma\Sigma h$ through $\mathbf{p}$ is contained in the set $|\Lambda^{s-1}_{h,\mathbf{z}}|$.*

Proof. Let C be an $(s-1)$-dimensional component of $\Sigma\Sigma h$ at $\mathbf{p}$. As all the Lê and polar numbers exist, we may inductively apply Proposition 1.21, together with Remark 1.10 and Proposition 1.16, to conclude that $\mathbf{p}$ is a singular point of the one-dimensional critical locus of $h_{|_{V(z_0-p_0,\dots,z_{s-2}-p_{s-2})}}$.

Using $(z_{s-1},\dots,z_n)$ as coordinates for $h_{|_{V(z_0-p_0,\dots,z_{s-2}-p_{s-2})}}$, we also conclude from Proposition 1.21 that, at $\mathbf{p}$,

$$\lambda^1_{h_{|_{V(z_0-p_0,\dots,z_{s-2}-p_{s-2})}}} \quad \text{and} \quad \gamma^1_{h_{|_{V(z_0-p_0,\dots,z_{s-2}-p_{s-2})}}}$$

exist – which, for a one-dimensional critical locus, is equivalent to $V(z_{s-1}-p_{s-1})$ being prepolar for $h_{|_{V(z_0-p_0,\dots,z_{s-2}-p_{s-2})}}$ at $\mathbf{p}$.

Therefore, by Proposition 1.30, $\lambda^0_{h_{|_{V(z_0-p_0,\dots,z_{s-2}-p_{s-2})}}}(\mathbf{p}) \neq 0$ or, equivalently, $\mathbf{p} \in \Gamma^1_{h_{|_{V(z_0-p_0,\dots,z_{s-2}-p_{s-2})}}}$. Now, by applying Proposition 1.21 once more, we find that $\mathbf{p} \in \Gamma^s_{h,\mathbf{z}}$.

As we may apply this same argument at each point of C near $\mathbf{p}$, we find that $C \subseteq \Gamma^s_{h,\mathbf{z}} \cap \Sigma h$. Finally, as the Lê numbers are defined, each of the Lê cycles has correct dimension at $\mathbf{p}$ and so the result follows from Proposition 1.15. □

Proposition 1.33. *Let $s = \dim_{\mathbf{p}}\Sigma h$, suppose that $\lambda^j_{h,\mathbf{z}}(\mathbf{p})$ and $\gamma^j_{h,\mathbf{z}}(\mathbf{p})$ exist for all $j \leqslant s$, and suppose that $\lambda^{s-1}_{h,\mathbf{z}}(\mathbf{p}) = 0$. Then, $\lambda^j_{h,\mathbf{z}}(\mathbf{p}) = 0$ for all $j \leqslant s-1$.*

Proof. The result follows from Proposition 1.31, using $i = s-1$, since the preceding proposition proves: if there exist two irreducible components, V and W, of Σh at $\mathbf{p}$ such that $\dim_{\mathbf{p}}(V \cap W) = s-1$, then $\mathbf{p} \in \Lambda^{s-1}_{h,\mathbf{z}}$ and so $\lambda^{s-1}_{h,\mathbf{z}}(\mathbf{p}) \neq 0$. □

Chapter 2. ELEMENTARY EXAMPLES

Example 2.1. If $\mathbf{0}$ is an isolated singularity of h, then regardless of the coordinate system $\mathbf{z}$, it follows from Proposition 1.11 that the only possibly non-zero Lê number is $\lambda^0_{h,\mathbf{z}}(\mathbf{0})$. Moreover, as $V\left(\frac{\partial h}{\partial z_0}, \frac{\partial h}{\partial z_1}, \dots, \frac{\partial h}{\partial z_n}\right)$ is zero-dimensional, $V\left(\frac{\partial h}{\partial z_1}, \dots, \frac{\partial h}{\partial z_n}\right)$ is one-dimensional with no components contained in Σh and with no embedded components. Therefore,

$$\Gamma^1_{h,\mathbf{z}} = V\left(\frac{\partial h}{\partial z_1}, \dots, \frac{\partial h}{\partial z_n}\right)$$

and so

$$\lambda^0_{h,\mathbf{z}}(\mathbf{0}) = \left(\Gamma^1_{h,\mathbf{z}} \cdot V\left(\frac{\partial h}{\partial z_0}\right)\right)_{\mathbf{0}} = \left(V\left(\frac{\partial h}{\partial z_1}, \dots, \frac{\partial h}{\partial z_n}\right) \cdot V\left(\frac{\partial h}{\partial z_0}\right)\right)_{\mathbf{0}} =$$

$$\left[V\left(\frac{\partial h}{\partial z_0}, \dots, \frac{\partial h}{\partial z_n}\right)\right]_{\mathbf{0}} = \text{the Milnor number of } h \text{ at } \mathbf{0}.$$

Example 2.2. Here, we generalize Example 1.12. Let $h = y^2 - x^a - tx^b$, where $a > b > 1$. We fix the coordinate system $\mathbf{z} = (t, x, y)$ and will suppress any further reference to it.

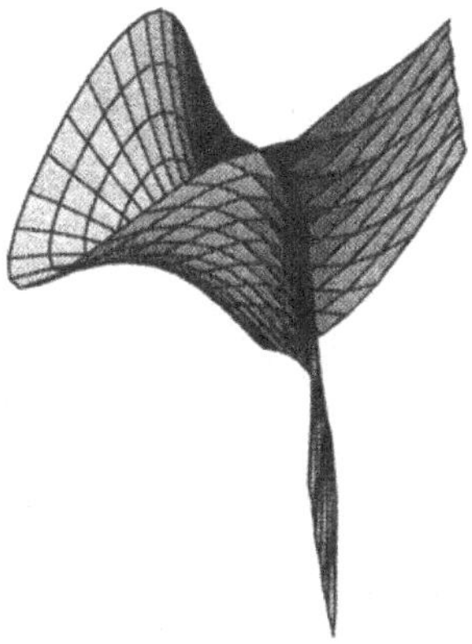

Figure 2.3. Generalization of nodes degenerating to a cusp

We find:

$$\Sigma h = V(-x^b,\ -ax^{a-1} - btx^{b-1},\ 2y) = V(x, y).$$

$$\Gamma_h^2 = V\left(\frac{\partial h}{\partial y}\right) = V(2y) = V(y).$$

$$\Gamma_h^2 \cdot V\left(\frac{\partial h}{\partial x}\right) = V(y) \cdot V(-ax^{a-1} - btx^{b-1}) =$$

$$V(y) \cdot (V(-ax^{a-b} - bt) + V(x^{b-1})) = V(-ax^{a-b} - bt, y) + (b-1)V(x,y)$$

$$= \Gamma_h^1 + \Lambda_h^1.$$

$$\Gamma_h^1 \cdot V\left(\frac{\partial h}{\partial t}\right) = V(-ax^{a-b} - bt, y) \cdot V(-x^b) = bV(t,x,y) = b[\mathbf{0}] = \Lambda_h^0.$$

Thus, $\lambda_h^0(\mathbf{0}) = b$ and $\lambda_h^1(\mathbf{0}) = b - 1$.

Notice that the exponent a does not appear; this is because $h = y^2 - x^a - tx^b = y^2 - x^b(x^{a-b} - t)$ which, after an analytic coordinate change at the origin, equals $y^2 - x^b u$.

Example 2.4 (The FM Cone). Let $h = y^2 - x^3 - (u^2 + v^2 + w^2)x^2$ and fix the coordinates (u, v, w, x, y).

$$\Sigma h = V(-2ux^2,\ -2vx^2,\ -2wx^2,\ -3x^2 - 2x(u^2 + v^2 + w^2),\ 2y) = V(x,y).$$

As Σh is three-dimensional, we begin our calculation with Γ_h^4.

$$\Gamma_h^4 = V(-2y) = V(y).$$

$$\Gamma_h^4 \cdot V\left(\frac{\partial h}{\partial x}\right) = V(y) \cdot V(-3x^2 - 2x(u^2 + v^2 + w^2)) =$$

$$V(-3x - 2(u^2 + v^2 + w^2), y) + V(x,y) = \Gamma_h^3 + \Lambda_h^3.$$

$$\Gamma_h^3 \cdot V\left(\frac{\partial h}{\partial w}\right) = V(-3x - 2(u^2 + v^2 + w^2), y) \cdot V(-2wx^2) =$$

$$V(-3x - 2(u^2 + v^2), w, y) + 2V(u^2 + v^2 + w^2, x, y) = \Gamma_h^2 + \Lambda_h^2.$$

$$\Gamma_h^2 \cdot V\left(\frac{\partial h}{\partial v}\right) = V(-3x - 2(u^2 + v^2), w, y) \cdot V(-2vx^2) =$$

$$V(-3x - 2u^2, v, w, y) + 2V(u^2 + v^2, w, x, y) = \Gamma_h^1 + \Lambda_h^1.$$

$$\Gamma_h^1 \cdot V\left(\frac{\partial h}{\partial u}\right) = V(-3x - 2u^2, v, w, y) \cdot V(-2ux^2) =$$

$$V(u,v,w,x,y) + 2V(u^2,v,w,x,y) = 5[\mathbf{0}] = \Lambda_h^0.$$

Hence, $\Lambda_h^3 = V(x,y)$, $\Lambda_h^2 = 2V(u^2+v^2+w^2,x,y)$ = a cone (as a set), $\Lambda_h^1 = 2V(u^2+v^2,w,x,y)$, and $\Lambda_h^0 = 5[0]$. Thus, at the origin, $\lambda_h^3 = 1$, $\lambda_h^2 = 4$, $\lambda_h^1 = 4$, and $\lambda_h^0 = 5$.

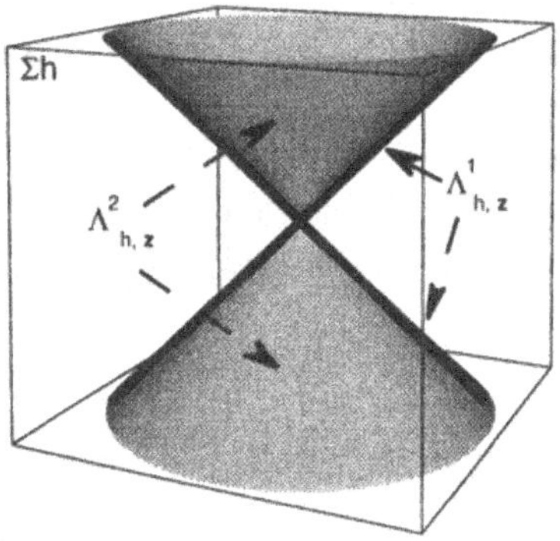

Figure 2.5. The critical locus of h

Note that Λ_h^1 depends on the choice of coordinates – for, by symmetry, if we re-ordered u, v, and w, then Λ_h^1 would change correspondingly. Moreover, one can check that this is a generic problem.

Such "non-fixed" Lê cycles arise from the absolute polar varieties (see [**L-T2**], [**Te2**], [**Te3**]) of the higher dimensional Lê cycles (we shall see this in Chapter 10, Remark 10.16). For instance, in the present case, Λ_h^2 is a cone, and its one-dimensional polar variety varies with the choice of coordinates, but generically always consists of two lines; this is the case for Λ_h^1 as well. Though the Lê cycles are not even generically fixed, the Lê numbers turn out to be generically independent of the coordinates (see Chapter 10, Theorem 10.18).

Example 2.6. Let $h = xyz$, so that $V(h)$ consists of the coordinate planes in $\mathbb{C}^3$. (See Example 0.2.) Then, $\Sigma h = V(x,y) \cup V(x,z) \cup V(y,z)$ = union of the three coordinate axes.

The coordinates (x,y,z) are extremely non-generic, so choose some other generic coordinates $\tilde{\mathbf{z}} = (\tilde{z}_0, \tilde{z}_1, \tilde{z}_2)$. Then, the set $|\Lambda^1_{h,\tilde{\mathbf{z}}}| = \Sigma h$. Hence,

$$\lambda^1_{h,\tilde{\mathbf{z}}}(\mathbf{0}) = \left(\Lambda^1_{h,\tilde{\mathbf{z}}} \cdot V(\tilde{z}_0)\right)_{\mathbf{0}} = \sum_{\mathbf{p}} \left(\Lambda^1_{h,\tilde{\mathbf{z}}} \cdot V(\tilde{z}_0 - \xi)\right)_{\mathbf{p}} = \sum_{\mathbf{p}} \lambda^1_{h,\tilde{\mathbf{z}}}(\mathbf{p}),$$

where the sum is over all $\mathbf{p} \in \overset{\circ}{B}_\epsilon \cap \Lambda^1_{h,\tilde{\mathbf{z}}} \cap V(\tilde{z}_0 - \xi)$ for small ϵ and $0 < \xi \ll \epsilon$;

this set consists of three points and, by symmetry, λ^1 must be the same at each of these three points. We wish to use Proposition 1.26 to calculate $\lambda^1_{h,\tilde{\mathbf{z}}}(\mathbf{p})$.

As each $\mathbf{p} \in \Sigma h$, it follows from 1.10 that $\gamma^1_{h,\tilde{\mathbf{z}}}$ is supported only at $\Lambda^0_{h,\tilde{\mathbf{z}}}$, which is generically zero-dimensional. Thus, our points $\mathbf{p}$ are such that $\gamma^1_{h,\tilde{\mathbf{z}}}(\mathbf{p}) = 0$, and it follows from 1.26 that $\lambda^1_{h,\tilde{\mathbf{z}}}(p) = \lambda^0_{h_{|V(\tilde{z}_0 - p_0)}, \hat{\mathbf{z}}}(\mathbf{p})$, where $\hat{\mathbf{z}}$ denotes the restriction of the coordinates $\tilde{\mathbf{z}}$ to $V(\tilde{z}_0 - p_0)$.

Now, $h_{|V(\tilde{z}_0 - p_0)}$ has an isolated singularity at each of our three points $\mathbf{p}$, and so $\lambda^0_{h_{|V(\tilde{z}_0 - p_0)}, \hat{\mathbf{z}}}(\mathbf{p})$ = the Milnor number of $h_{|V(\tilde{z}_0 - p_0)}$ at $\mathbf{p}$, and this is easily seen to equal 1. It follows, finally, that $\lambda^1_{h,\tilde{\mathbf{z}}}(\mathbf{0}) = 3$.

The generic value of $\lambda^0_{h,\tilde{\mathbf{z}}}(\mathbf{0})$ is somewhat messier to calculate, and is just as easy to treat in the more general case given in Example 2.8. (However, the answer is that $\lambda^0_{h,\tilde{\mathbf{z}}}(\mathbf{0}) = 2$.)

Example 2.7. Let $\mathcal{U}$ be an open subset of $\mathbb{C}^{n+1}$, let $\mathbf{z} = (z_0, \dots, z_n)$ be the coordinates for $\mathbb{C}^{n+1}$, and $h : \mathcal{U} \to \mathbb{C}$ be any analytic function. The coordinates $\mathbf{z}$ may be non-generic for h. We wish to see how to calculate $\lambda^0_{h,\tilde{\mathbf{z}}}$ for a generic linear choice of $\tilde{\mathbf{z}}$.

So, let $\tilde{\mathbf{z}}$ be a generic linear choice of coordinates for $\mathbb{C}^{n+1}$, and let a_{ij} denote $\frac{\partial z_i}{\partial \tilde{z}_j}$.

Now,

$$\Gamma^1_{h,\tilde{\mathbf{z}}} = V\left(\frac{\partial h}{\partial \tilde{z}_1}, \dots, \frac{\partial h}{\partial \tilde{z}_n}\right) \Big/ \Sigma h =$$

$$V\left(a_{01}\frac{\partial h}{\partial z_0} + \dots + a_{n1}\frac{\partial h}{\partial z_n}, \ \dots \ , \ a_{0n}\frac{\partial h}{\partial z_0} + \dots + a_{nn}\frac{\partial h}{\partial z_n}\right) / \Sigma h.$$

By performing elementary row operations, we find that the ideal

$$\left\langle a_{01}\frac{\partial h}{\partial z_0} + \dots + a_{n1}\frac{\partial h}{\partial z_n}, \ \dots \ , \ a_{0n}\frac{\partial h}{\partial z_0} + \dots + a_{nn}\frac{\partial h}{\partial z_n}\right\rangle$$

is generated by

$$\frac{\partial h}{\partial z_0} + b_0\frac{\partial h}{\partial z_n}, \ \frac{\partial h}{\partial z_1} + b_1\frac{\partial h}{\partial z_n}, \ \dots, \ \frac{\partial h}{\partial z_{n-1}} + b_{n-1}\frac{\partial h}{\partial z_n},$$

where $b_0, \dots, b_{n-1}$ are generic $\neq 0$.

Thus,

$$\Gamma^1_{h,\tilde{\mathbf{z}}} = V\left(\frac{\partial h}{\partial z_0} + b_0\frac{\partial h}{\partial z_n}, \frac{\partial h}{\partial z_1} + b_1\frac{\partial h}{\partial z_n}, \dots, \frac{\partial h}{\partial z_{n-1}} + b_{n-1}\frac{\partial h}{\partial z_n}\right) / \Sigma h,$$

and $\Lambda^0_{h,\tilde{\mathbf{z}}}$ is given by intersecting this with $a_{00}\frac{\partial h}{\partial z_0} + \dots + a_{n0}\frac{\partial h}{\partial z_n}$.

It is important to note that we are **not** claiming that the cycle $\Gamma^1_{h,\tilde{\mathbf{z}}}$ can be calculated by considering the cycle

$$V\left(\frac{\partial h}{\partial z_0}+b_0\frac{\partial h}{\partial z_n}\right)\cdot V\left(\frac{\partial h}{\partial z_1}+b_1\frac{\partial h}{\partial z_n}\right)\cdot\ldots\cdot V\left(\frac{\partial h}{\partial z_{n-1}}+b_{n-1}\frac{\partial h}{\partial z_n}\right)$$

and then disposing of any portions of the cycle which are contained in Σh. There could easily be a problem with embedded subvarieties.

Example 2.8. We can use the above example to calculate the generic value of λ^0 in Example 2.6. Actually, we can just as easily do a more general calculation.

Let $h = z_0z_1\ldots z_n$, so that $V(h)$ is the union of the coordinate planes in $\mathbb{C}^{n+1}$ and $\Sigma h = \bigcup_{i\neq j} V(z_i, z_j)$ = the union of intersections of pairs of the different coordinate planes. We wish to show, for a generic choice of coordinates, $\tilde{\mathbf{z}}$, that $\lambda^0_{h,\tilde{\mathbf{z}}}(\mathbf{0}) = n$.

By the above, we find that $\Gamma^1_{h,\mathbf{z}}$ equals

$$V\big(z_1z_2\ldots z_{n-1}(z_n+b_0z_0),\ z_0z_2\ldots z_{n-1}(z_n+b_1z_1),\ \ldots,\ z_0z_1\ldots z_{n-2}(z_n+b_{n-1}z_{n-1})\big)/\Sigma h.$$

Applying 1.2.iii repeatedly, we conclude that

$$\Gamma^1_{h,\mathbf{z}} = V(z_n+b_0z_0,\ z_n+b_1z_1,\ \ldots,\ z_n+b_{n-1}z_{n-1}).$$

Finally, by intersecting this with

$$V\left(\frac{\partial h}{\partial \tilde{z}_0}\right) = V(a_{00}z_1z_2\ldots z_n+\cdots+a_{n0}z_0z_1\ldots z_{n-1})$$

we obtain the desired result that $\lambda^0_{h,\tilde{\mathbf{z}}}(\mathbf{0}) = n$.

We shall obtain this same result, but by inductive methods, in Example 5.2.

Example 2.9. Let h be an analytic map in the variables x and y, and suppose that $h = P\prod Q_i^{\alpha_i}$, where P and $\prod Q_i^{\alpha_i}$ are relatively prime and $\alpha_i \geqslant 2$, i.e. h gives a non-reduced curve singularity. We wish to calculate the Lê numbers of h at the origin.

Let $z_0 = ax+by$, where $a\neq 0$, and let $z_1 = y$. Then,

$$\left[V\left(\frac{\partial h}{\partial z_1}\right)\right] = \left[V\left(\frac{\partial h}{\partial x}\left(\frac{-b}{a}\right)+\frac{\partial h}{\partial y}\right)\right] =$$

$$\sum\left[V\left(Q_i^{\alpha_i-1}\right)\right] + \left[V\left(\frac{\frac{\partial h}{\partial x}\left(\frac{-b}{a}\right)+\frac{\partial h}{\partial y}}{\prod Q_i^{\alpha_i-1}}\right)\right] = \Lambda^1_{h,\mathbf{z}} + \Gamma^1_{h,\mathbf{z}}.$$

Thus, whenever
$$\left[V\left(\frac{\frac{\partial h}{\partial x}\left(\frac{-b}{a}\right)+\frac{\partial h}{\partial y}}{\prod Q_i^{\alpha_i-1}}\right)\right]$$
has no components contained in the critical locus of h (an easy argument shows that this is the case for a generic choice of (a, b)), we have that
$$\lambda^1_{h,\mathbf{z}}=\sum(\alpha_i-1)\,(V(Q_i)\cdot V(ax+by))_{\mathbf{0}}$$
and
$$\lambda^0_{h,\mathbf{z}}=\left(V\left(\frac{\frac{\partial h}{\partial x}\left(\frac{-b}{a}\right)+\frac{\partial h}{\partial y}}{\prod Q_i^{\alpha_i-1}}\right)\cdot V\left(\frac{\partial h}{\partial x}\right)\right)_{\mathbf{0}},$$
where we have used that $V\left(\frac{\partial h}{\partial z_0}\right)=V\left(\frac{\partial h}{\partial x}\right)$.

Note that the formula
$$\lambda^1_{h,\mathbf{z}}=\sum(\alpha_i-1)\,(V(Q_i)\cdot V(ax+by))_{\mathbf{0}}$$
agrees with our earlier formula from the end of Remark 1.19,
$$\lambda^1_{h,\mathbf{z}}=\sum_{\nu}n_\nu\mathring{\mu}_\nu,$$
since we clearly have $n_\nu=(V(Q_i)\cdot V(ax+by))_{\mathbf{0}}$ and $\mathring{\mu}_\nu=\alpha_i-1$.

Example 2.10. In this example, we show that – unlike the Milnor number – the Lê numbers in a family need **not** be upper-semicontinuous. While this may seem to be mildly disturbing at first, the example makes it clear what can happen; if a high-dimensional Lê number jumps up, then the lower-dimensional Lê numbers are free to jump up or down.

Let $f_t(x,y,z)=z^2-y^3-txy^2$. The coordinates (x,y,z) are prepolar at the origin for f_t for all t; we fix this set of coordinates and will suppress further reference to them.

For $t_0\neq 0$, we are back in the situation of Example 2.2, with $a=3$ and $b=2$; therefore, $\lambda^0_{f_{t_0}}(\mathbf{0})=2$ and $\lambda^1_{f_{t_0}}(\mathbf{0})=1$.

On the other hand, the hypersurface defined by f_0 is a cross-product singularity; hence, $\lambda^0_{f_0}(\mathbf{0})=0$, and one trivially finds that $\lambda^1_{f_0}(\mathbf{0})=2$.

Thus, at $t=0$, λ^1 jumps up to 2 from its generic value of 1; this allows the behavior of $\lambda^0_{f_t}(\mathbf{0})$ to be about as "bad" as possible; the generic value of λ^0 is 2, while the special value is 0.

The situation is not completely uncontrolled – as we shall see in Corollary 4.16, if we have a family f_t, then the tuple of Lê numbers
$$\left(\lambda^s_{f_t,\mathbf{z}}(\mathbf{0}),\lambda^{s-1}_{f_t,\mathbf{z}}(\mathbf{0}),\dots,\lambda^0_{f_t,\mathbf{z}}(\mathbf{0})\right)$$
is lexigraphically upper-semicontinuous in the t variable.

Chapter 3. A HANDLE DECOMPOSITION OF THE MILNOR FIBRE

In this chapter, we give a handle decomposition of the Milnor fibre of an analytic function with a critical locus of arbitrary dimension. This decomposition is more refined than that obtained by iteratively applying Lê's attaching result (Theorem 0.9).

Throughout this chapter, $h : \mathcal{U} \to \mathbb{C}$ will be an analytic function on an open subset of $\mathbb{C}^{n+1}$ and, if $\mathbf{p} \in V(h)$, then we let $F_{h,\mathbf{p}}$ denote the Milnor fibre of h at $\mathbf{p}$.

Our main tool is a proposition based on the argument of Lê and Perron [**L-P**], which is the same argument that is used in [**Ti**], [**Va1**], and [**Va2**]. In what follows, if we have a pair of topological spaces $X \subseteq Y$, then we say that Y *is obtained from* X *by cancelling* k m*-handles* provided that X has a handle decomposition in which the handles of highest index are of index m and Y is obtained from X by attaching k $(m+1)$-handles each of which cancels with an m-handle of X (in terms of Morse functions, this says that Y is obtained from X by passing through k critical points – all of index $m+1$ – and these critical points cancel with k critical points in X each of index m. See [**Mi1**] and [**Sm**].) In particular, this implies that the cohomology groups of X and Y are identical except in degree m, where we have $H^m(X) \cong \mathbb{Z}^k \oplus H^m(Y)$.

Proposition 3.1. *Let* $\mathbf{p} \in V(h)$. *If* $V(z_0 - p_0)$ *is a prepolar slice for* h *at* $\mathbf{p}$, *and* $n \neq 2$, *then the Milnor fibre of* h *at* $\mathbf{p}$ *is obtained – up to diffeomorphism – from the product of a disk with the Milnor fibre of* $h_{|_{V(z_0-p_0)}}$ *at* $\mathbf{p}$ *by first attaching* $\gamma^1_{h,\mathbf{z}}(\mathbf{p})$ n*-handles, which cancel against* $\gamma^1_{h,\mathbf{z}}(\mathbf{p})$ $(n-1)$*-handles of* $\overset{\circ}{\mathbb{D}} \times F_{h_{|_{V(z_0-p_0)}},\mathbf{p}}$, *and then attaching* $\lambda^0_{h,\mathbf{z}}(\mathbf{p})$ *more* n*-handles.*

If $n = 2$, *we have the same conclusion except that the cancelling is only up to homotopy.*

Proof. Essentially, this is Proposition 4.2 of [**Mas2**], except that here we have weakened the hypothesis on the genericity of the hyperplane slice. We use the coodinates $(z_0, \ldots, z_n)$ for our ambient space. Clearly, it suffices to prove the claim for $\mathbf{p} = \mathbf{0}$. We will follow the argument of [**Va2**].

By Proposition A.6.iii, we may use neighborhoods of the form $\mathbb{D}_\delta \times B_\epsilon$, $0 < \delta \ll \epsilon$, to define the Milnor fibre of h at the origin up to homotopy. Choose ϵ and δ such that B_ϵ is a Milnor ball for $h_{|_{V(z_0)}}$ and $\Gamma^1_{h,z_0} \cap (\mathbb{D}_\delta \times \partial B_\epsilon) = \emptyset$ – we may accomplish this last equality since Theorem 1.28 implies that $\dim_{\mathbf{0}}(\Gamma^1_{h,z_0} \cap V(z_0)) \leqslant 0$. Choose η such that $(B_\epsilon, \mathbb{D}_\eta)$ is a Milnor pair for $h_{|_{V(z_0)}}$ at the origin (see the appendix). Let $\Psi := (h, z_0)$ and let Δ denote $\Psi(\Gamma^1_{h,z_0})$ in $\mathbb{C}^2$ (Δ is the *Cerf diagram* of h with respect to z_0). Δ is given its fitting ideal structure, which is possibly non-reduced (see [Lo]).

Choose $(\alpha, \beta) \in \mathbb{C}^2 - \Delta$ sufficiently small and let $h_\beta := h_{|_{V(z_0-\beta)}}$. Let $\mathbb{D}$ be a small disc in $\mathbb{D}_\eta \times \{\beta\}$ centered at (α, β), and let A be the region in $\mathbb{D}_\eta \times \{\beta\}$ formed by joining to $\mathbb{D}$ small discs centered at each of the points of $\Delta \cap V(z_0 - \beta)$, where the joining is via thickened paths which avoid $(0, \beta)$ (see Figure 3.2). Note that, counted with multiplicity, there are $(\Gamma^1_{h,z_0} \cdot V(z_0))_\mathbf{0}$ points in $\Delta \cap V(z_0 - \beta)$.

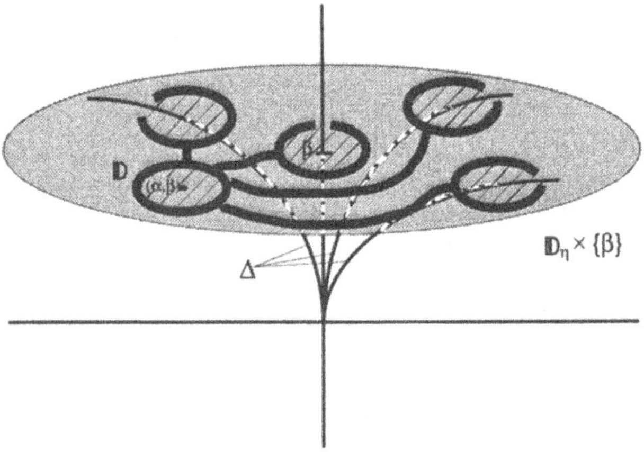

Figure 3.2. The Cerf diagram, and the sets A and C

Then, the argument of Lê and Perron [L-P] and Vannier [Va1] shows that the Milnor fibre of h is obtained from $W := h_\beta^{-1}(A) \cap (\{\beta\} \times B_\epsilon)$ by attaching $\lambda^0_{h,z_0}(\mathbf{0}) = \left(\Gamma^1_{h,z_0} \cdot V\left(\frac{\partial h}{\partial z_0}\right)\right)_\mathbf{0}$ n-handles – this part of their argument does not depend on the dimension of Σh. (Though, as we are not assuming that the polar curve is reduced, the details of the generalization of the isotopy result used by Lê and Perron need to be checked – this is done in [Ti].)

The problem is thus to show that W is obtained from the product of a disc with the Milnor fibre of $h_{|_{V(z_0)}}$ by cancelling $\gamma^1_{h,z_0}(\mathbf{0}) = (\Gamma^1_{h,z_0} \cdot V(z_0))_\mathbf{0}$ $(n-1)$-handles, even in the case where the critical locus of h does not have dimension one.

Let $C \subseteq \mathbb{D}_\eta \times \{\beta\}$ be formed by taking a small disc around $(0, \beta)$ and joining it to $\mathbb{D}$ with a thickened path (see Figure 3.2). Let $U := h_\beta^{-1}(C) \cap (\{\beta\} \times B_\epsilon)$. Then, as $A \cup C$ is a strong deformation retract of $\mathbb{D}_\eta$, $U \cup V$ is homotopy-equivalent to $h_\beta^{-1}(\mathbb{D}_\eta) \cap (\{\beta\} \times B_\epsilon)$, which is in turn diffeomorphic to a real $2n$-ball, B^{2n}. Moreover, $U \cap W$ is diffeomorphic to $h_\beta^{-1}(\mathbb{D})$ which is diffeomorphic to the product of a disc and the Milnor fibre of $h_{|_{V(z_0)}}$.

Now, $U \cup W$ is obtained from U by attaching $\gamma^1_{h,z_0}(\mathbf{0})$ n-handles (that is, one handle of index n for each point of $\Delta \cap V(z_0 - \beta)$ in A, counted with multiplicity). As $U \cup W$ is contractible, this implies that U has the homotopy-type of a bouquet of $\gamma^1_{h,z_0}(\mathbf{0})$ $(n-1)$-spheres.

Consider the Mayer-Vietoris sequence of U and W. As $U \cup W$ is contractible, we have that $H_k(U \cap W) \cong H_k(U) \oplus H_k(W)$ for all $k \geqslant 1$, where we know that U has the homotopy-type of a bouquet of $(n-1)$-spheres, and $U \cap W$ is diffeomorphic to the product of a disc and the Milnor fibre of $h_{|V(z_0)}$, which has the homotopy-type of an $(n-1)$-dimensional CW complex. Thus, we see that W has the homotopy-type of an $(n-1)$-dimensional CW complex.

But now, since $h_\beta^{-1}(\mathbb{D})$ is diffeomorphic to the product of a disc with the Milnor fibre of $h_{|V(z_0)}$, and since W itself is obtained by attaching $\gamma^1_{h,z_0}(\mathbf{0})$ n-handles to $h_\beta^{-1}(\mathbb{D})$, we see that up to homotopy these $\gamma^1_{h,z_0}(\mathbf{0})$ n-handles must be cancelling $\gamma^1_{h,z_0}(\mathbf{0})$ $(n-1)$-handles. But, if $n \geqslant 3$, then the real dimension of W is greater than or equal to 6, and so this handle cancellation is up to diffeomorphism [**Mi1**], [**Sm**].

Finally, if $n = 1$, then – by the classification of surfaces – we have that the handle cancellation is up to diffeomorphism. □

By an inductive application of Proposition 3.1 to each hyperplane slice, we arrive at Theorem 4.3 of [**Mas2**]. This theorem describes a handle decomposition of the Milnor fibre.

Theorem 3.3. *Let $\mathcal{U}$ be an open subset of $\mathbb{C}^{n+1}$, let $h : \mathcal{U} \to \mathbb{C}$ be an analytic map, let $\mathbf{p} \in V(h)$, let s denote $\dim_{\mathbf{p}} \Sigma h$, and let $\mathbf{z} = (z_0, \ldots, z_{s-1})$ be prepolar for h at $\mathbf{p}$.*

If $s \leqslant n-2$, then $F_{h,\mathbf{p}}$ is obtained up to diffeomorphism from a real $2n$-ball by successively attaching $\lambda^{n-k}_{h,\mathbf{z}}(\mathbf{p})$ k-handles, where $n-s \leqslant k \leqslant n$;

if $s = n-1$, then $F_{h,\mathbf{p}}$ is obtained up to diffeomorphism from a real $2n$-manifold with the homotopy-type of a bouquet of $\lambda^{n-1}_{h,\mathbf{z}}(\mathbf{p})$ circles by successively attaching $\lambda^{n-k}_{h,\mathbf{z}}(\mathbf{p})$ k-handles, where $2 \leqslant k \leqslant n$

Hence, the reduced Euler characteristic of the Milnor fibre of h at $\mathbf{p}$ is given by

$$\widetilde{\chi}(F_{h,\mathbf{p}}) = \sum_{i=0}^{s} (-1)^{n-i} \lambda^i_{h,\mathbf{z}}(\mathbf{p})$$

and the reduced Betti numbers, $\tilde{b}_i(F_{h,\mathbf{p}})$, satisfy Morse inequalities with respect to the Lê numbers, i.e. for all k with $n-s \leqslant k \leqslant n$,

$$(-1)^k \sum_{i=n-s}^{k} (-1)^i \tilde{b}_i(F_{h,\mathbf{p}}) \leqslant (-1)^k \sum_{i=n-s}^{k} (-1)^i \lambda^{n-i}_{h,\mathbf{z}}(\mathbf{p})$$

and

$$(-1)^k \sum_{i=k}^{n} (-1)^i \tilde{b}_i(F_{h,\mathbf{p}}) \leqslant (-1)^k \sum_{i=k}^{n} (-1)^i \lambda^{n-i}_{h,\mathbf{z}}(\mathbf{p})$$

Proof. By induction on s. When $s = 0$, the result follows from Milnor's work. Now, assume the result for $s - 1$. As before, we consider only the case where $\mathbf{p} = \mathbf{0}$.

As $V(z_0)$ is prepolar, we may apply Proposition 3.1 to conclude that the Milnor fibre of h at $\mathbf{0}$ is obtained from the product of a disk with the Milnor fibre of $h_{|_{V(z_0)}}$ at $\mathbf{0}$ by first attaching $\gamma^1_{h,\mathbf{z}}(\mathbf{0})$ n-handles, which cancel against $\gamma^1_{h,\mathbf{z}}(\mathbf{0})$ $(n-1)$-handles of $\overset{\circ}{\mathbb{D}} \times F_{h_{|_{V(z_0)}},\mathbf{0}}$, and then attaching $\lambda^0_{h,\mathbf{z}}(\mathbf{0})$ more n-handles – up to diffeomorphism if $n \neq 2$ and up to homotopy otherwise.

But, $\tilde{\mathbf{z}} = (z_1, \dots, z_{s-1})$ is prepolar for $h_{|_{V(z_0)}}$ at the origin and, by our inductive hypothesis, the Milnor fibre of $h_{|_{V(z_0)}}$ at the origin is obtained by successively attaching $\lambda^{n-1-k}_{h_{|_{V(z_0)}},\tilde{\mathbf{z}}}(\mathbf{0})$ k-handles for $(n-1)-(s-1) \leqslant k \leqslant n-1$. By Proposition 1.8, if $n - 1 - k \neq 0$, then $\lambda^{n-1-k}_{h_{|_{V(z_0)}},\tilde{\mathbf{z}}}(\mathbf{0}) = \lambda^{n-k}_{h,\mathbf{z}}(\mathbf{0})$, and $\lambda^0_{h_{|_{V(z_0)}},\tilde{\mathbf{z}}}(\mathbf{0}) = \gamma^1_{h,\mathbf{z}}(\mathbf{0}) + \lambda^1_{h,\mathbf{z}}(\mathbf{0})$.

The conclusion follows. □

Siersma's main result in [**Si**] allows us to improve this result in a special case.

Corollary 3.4 *Let $\mathcal{U}$ be an open subset of $\mathbb{C}^{n+1}$, let $h : \mathcal{U} \to \mathbb{C}$ be an analytic map, let $\mathbf{p} \in V(h)$, and let s denote $\dim_{\mathbf{p}}\Sigma h$. Suppose that $(z_0, \dots, z_{s-1})$ is prepolar for h at $\mathbf{p}$, and suppose that $\lambda^s_{h,\mathbf{z}}(\mathbf{p}) = 1$.*

Then, either

$$\lambda^0_{h,\mathbf{z}}(\mathbf{p}) = \lambda^1_{h,\mathbf{z}}(\mathbf{p}) = \cdots = \lambda^{s-1}_{h,\mathbf{z}}(\mathbf{p}) = 0,$$

or the single $(n-s)$-handle in the handle decomposition of the previous theorem gets cancelled – up to homotopy – by the attaching of one of the $\lambda^{s-1}_{h,\mathbf{z}}(\mathbf{p})$ $(n-s+1)$-handles.

Proof. The proof is exactly the inductive proof of 3.3, except in the first step one applies the result of [**Si**].

The function $h_{|_{V(z_0-p_0,\dots,z_{s-2}-p_{s-2})}}$ has a one-dimensional critical locus at $\mathbf{p}$. Using $(z_{s-1}, \dots, z_n)$ as coordinates for $V(z_0 - p_0, \dots, z_{s-2} - p_{s-2})$, we conclude from 1.21 that

$$\lambda^1_{h_{|_{V(z_0-p_0,\dots,z_{s-2}-p_{s-2})}}}(\mathbf{p}) = \lambda^s_{h,\mathbf{z}}(\mathbf{p}) = 1$$

and

$$\lambda^0_{h_{|_{V(z_0-p_0,\dots,z_{s-2}-p_{s-2})}}}(\mathbf{p}) = \gamma^{s-1}_{h,\mathbf{z}}(\mathbf{p}) + \lambda^{s-1}_{h,\mathbf{z}}(\mathbf{p}).$$

Hence, $h_{|_{V(z_0-p_0,\dots,z_{s-2}-p_{s-2})}}$ is an isolated line singularity in the sense of Siersma [**Si**], and his result is that either $\lambda^0_{h_{|_{V(z_0-p_0,\dots,z_{s-2}-p_{s-2})}}}(\mathbf{p}) = 0$ or that one only has homology in middle dimension, i.e. the one possible $(n-s)$-handle must get cancelled up to homotopy.

The equality

$$\lambda^0_{h_{|V(z_0-p_0,\dots,z_{s-2}-p_{s-2})}}(\mathbf{p}) = \gamma^{s-1}_{h,\mathbf{z}}(\mathbf{p}) + \lambda^{s-1}_{h,\mathbf{z}}(\mathbf{p}) = 0$$

corresponds to the case where

$$\lambda^0_{h,\mathbf{z}}(\mathbf{p}) = \lambda^1_{h,\mathbf{z}}(\mathbf{p}) = \cdots = \lambda^{s-1}_{h,\mathbf{z}}(\mathbf{p}) = 0,$$

since $\lambda^{s-1}_{h,\mathbf{z}}(\mathbf{p}) = 0$ implies that $\mathbf{p} \notin \Gamma^{s-1}_{h,\mathbf{z}}$ and, by 1.15, $\Gamma^{s-1}_{h,\mathbf{z}} \cap \Sigma h = \bigcup_{i \leqslant s-2} \Lambda^i_{h,\mathbf{z}}$. Hence, if $\mathbf{p} \notin \Gamma^{s-1}_{h,\mathbf{z}}$, then it follows that all the lower Lê numbers are also zero. □

One might question whether the above result can possibly be correct. What about the case where $\lambda^s_{h,\mathbf{z}}(\mathbf{p}) = 1$, $\lambda^{s-1}_{h,\mathbf{z}}(\mathbf{p}) = 0$, and one of the lower λ's is not zero? In such a case, there would be no way to cancel the $(n-s)$-handle. Note, however, that 1.33 rules out the possibility of the existence of this case.

Chapter 4. GENERALIZED LÊ-IOMDINE FORMULAS

In this chapter, we generalize the formula of Lê and Iomdine (see [**Lê2**], [**Io**], [**Mas3**], [**Mas5**], [**M-S**], and [**Si2**]) to functions with an arbitrary-dimensional critical locus. This formula tells us how the Lê numbers of a hypersurface singularity are related to the Lê numbers of a certain "sequence of hypersurface singularities" – a sequence which "approaches" the original singularity, but such that the critical loci of the terms in the sequence are of one dimension smaller than the original. This formula has a large number of applications.

The statement that we give here has an improvement in a certain bound; in the one-dimensional case, this is the form of the statement as it appears in [**M-S**] and [**Si2**]. To give this improved bound, we need a definition. Throughout this chapter, we concentrate our attention at the origin.

Definition 4.1. Suppose that Γ^1_{h,z_0} is one-dimensional at the origin. Let η be an irreducible component of Γ^1_{h,z_0} (with its reduced structure) such that $\eta \cap V(z_0)$ is zero-dimensional at the origin.

Then, the *polar ratio* of η (for h at $\mathbf{0}$ with respect to z_0) is

$$\frac{(\eta \cdot V(h))_{\mathbf{0}}}{(\eta \cdot V(z_0))_{\mathbf{0}}} = \frac{\left(\eta \cdot V\left(\frac{\partial h}{\partial z_0}\right)\right)_{\mathbf{0}} + (\eta \cdot V(z_0))_{\mathbf{0}}}{(\eta \cdot V(z_0))_{\mathbf{0}}} = \frac{\left(\eta \cdot V\left(\frac{\partial h}{\partial z_0}\right)\right)_{\mathbf{0}}}{(\eta \cdot V(z_0))_{\mathbf{0}}} + 1.$$

(The equalities follow from our proof of Proposition 1.23.)

If $\eta \cap V(z_0)$ is not zero-dimensional at the origin (i.e. if $\eta \subseteq V(z_0)$), then we say that the polar ratio of η equals 1.

A *polar ratio* (of h at $\mathbf{0}$ with respect to z_0) is any one of the polar ratios of any component of the polar curve.

Remark 4.2. The case where h is a homogeneous polynomial of degree d is particularly easy to analyze. Provided that Γ^1_{h,z_0} is one-dimensional at the origin, each component of the polar curve is a line, and so the polar ratios are all 1 or d.

We are going to consider functions of the form $h + az_0^j$, where a is a non-zero complex number and j is suitably large. Clearly, however, the coordinate z_0 is extremely non-generic for $h + az_0^j$. Hence, if we are using the coordinates $(z_0, z_1, \dots, z_n)$ for h, we use the coordinates $(z_1, z_2, \dots, z_n, z_0)$ for $h + az_0^j$. The purpose of this "rotation" of the coordinate system is merely to get the z_0 coordinate out of the way. Normally, if h has an s-dimensional critical locus at the origin, then $h + az_0^j$ will have an $(s-1)$-dimensional critical locus at the origin; thus, it is only the choice of the coordinates $z_0, \ldots, z_{s-1}$ that we care about for h, and the coordinates $z_1, \ldots, z_{s-1}$ for $h + az_0^j$.

Lemma 4.3. *Let $j \geqslant 2$, let $h : (\mathcal{U}, \mathbf{0}) \to (\mathbb{C}, 0)$ be an analytic function, let s denote $\dim_{\mathbf{0}} \Sigma h$, and assume that $s \geqslant 1$. Let $\mathbf{z} = (z_0, \ldots, z_n)$ be a linear choice of coordinates such that $\lambda^i_{h,\mathbf{z}}(\mathbf{0})$ is defined for all $i \leqslant s$. Let a be a non-zero complex number, and use the coordinates $\tilde{\mathbf{z}} = (z_1, \ldots, z_n, z_0)$ for $h + az_0^j$.*

If j is greater than or equal to the maximum polar ratio for h then, for all but a finite number of complex a,

i) $\dim_{\mathbf{0}} \Gamma^1_{h,\mathbf{z}} \cap V\left(\frac{\partial h}{\partial z_0} + jaz_0^{j-1}\right) = 0$;

ii) $\lambda^0_{h,\mathbf{z}}(\mathbf{0}) = \left(\Gamma^1_{h,\mathbf{z}} \cdot V\left(\frac{\partial h}{\partial z_0} + jaz_0^{j-1}\right)\right)_{\mathbf{0}}$;

iii) *$\Sigma(h + az_0^j)$ is $(s-1)$-dimensional at the origin and equal to $\Sigma h \cap V(z_0)$ as germs of sets at $\mathbf{0}$;*

iv) *if $i \geqslant 1$, then we have an equality of cycles*

$$\Gamma^i_{h+az_0^j, \tilde{\mathbf{z}}} = \Gamma^{i+1}_{h,\mathbf{z}} \cdot V\left(\frac{\partial h}{\partial z_0} + jaz_0^{j-1}\right)$$

near the origin;

v) $\left(\Gamma^1_{h+aw^j, w-z_0} \cdot V(h + aw^j)\right)_{\mathbf{0}} = j\lambda^0_{h,\mathbf{z}}(\mathbf{0})$, *where w is a variable disjoint from those of h.*

Moreover, if we have the strict inequality that j is greater than the maximum polar ratio for h, then the above equalities hold for all non-zero a; in particular, this is the case if $j \geqslant 2 + \lambda^0_{h,\mathbf{z}}(\mathbf{0})$.

Proof.

Proof of i) and ii): As $\lambda^0_{h,\mathbf{z}}(\mathbf{0})$ is defined, $\Gamma^1_{h,\mathbf{z}}$ is one-dimensional at the origin. If we write the cycle $\Gamma^1_{h,\mathbf{z}}$ as $\sum_\eta k_\eta [\eta]$, where the η are the irreducible components of $\Gamma^1_{h,\mathbf{z}}$, then

$$\left(\Gamma^1_{h,\mathbf{z}} \cdot V\left(\frac{\partial h}{\partial z_0} + jaz_0^{j-1}\right)\right)_{\mathbf{0}} = \sum k_\eta \left(\eta \cdot V\left(\frac{\partial h}{\partial z_0} + jaz_0^{j-1}\right)\right)_{\mathbf{0}}.$$

It suffices to impose conditions on j and a so that each

$$\left(\eta \cdot V\left(\frac{\partial h}{\partial z_0} + jaz_0^{j-1}\right)\right)_{\mathbf{0}}$$

actually equals

$$\left(\eta \cdot V\left(\frac{\partial h}{\partial z_0}\right)\right)_{\mathbf{0}}$$

for then we would have

$$\dagger)\quad \left(\Gamma^1_{h,\mathbf{z}} \cdot V\left(\frac{\partial h}{\partial z_0} + jaz_0^{j-1}\right)\right)_{\mathbf{0}} = \left(\Gamma^1_{h,\mathbf{z}} \cdot V\left(\frac{\partial h}{\partial z_0}\right)\right)_{\mathbf{0}} = \lambda^0_h(\mathbf{0}).$$

By [**Fu**], if we let $\alpha_\eta(t)$ be a parametrization of η, we may calculate the intersection number

$$\left(\eta \cdot V\left(\frac{\partial h}{\partial z_0} + jaz_0^{j-1}\right)\right)_{\mathbf{0}}$$

by taking the t-multiplicity of $\left(\frac{\partial h}{\partial z_0} + jaz_0^{j-1}\right)|_{\alpha_\eta(t)}$.

Using this same method of calculating the intersection number twice more, we conclude that

$$\left(\eta \cdot V\left(\frac{\partial h}{\partial z_0} + jaz_0^{j-1}\right)\right)_{\mathbf{0}} =$$

$$\min\left\{\left(\eta \cdot V\left(\frac{\partial h}{\partial z_0}\right)\right)_{\mathbf{0}}, \left(\eta \cdot V\left(z_0^{j-1}\right)\right)_{\mathbf{0}}\right\}$$

with the exception of, possibly, the single value of a which makes the lowest degree terms of $\left(\frac{\partial h}{\partial z_0}\right)|_{\alpha_\eta(t)}$ and $\left(jaz_0^{j-1}\right)|_{\alpha_\eta(t)}$ add up to zero.

Thus, using that $\left(\eta \cdot V\left(z_0^{j-1}\right)\right)_{\mathbf{0}} = (j-1)\left(\eta \cdot V(z_0)\right)_{\mathbf{0}}$, we find that if we have

$$*)\quad j \geqslant \frac{\left(\eta \cdot V\left(\frac{\partial h}{\partial z_0}\right)\right)_{\mathbf{0}}}{\left(\eta \cdot V(z_0)\right)_{\mathbf{0}}} + 1$$

for all η, then †) holds for all but a finite number of a. In addition, if we choose strict inequalities in $*$), then †) holds for every value of a. This proves i) and ii).

Proof of iii): As sets,

$$\Sigma(h + az_0^j) = V\left(\frac{\partial h}{\partial z_0} + jaz_0^{j-1}, \frac{\partial h}{\partial z_1}, \dots, \frac{\partial h}{\partial z_n}\right) =$$

$$V\left(\frac{\partial h}{\partial z_0} + jaz_0^{j-1}\right) \cap \left(\Gamma^1_{h,\mathbf{z}} \cup \Sigma h\right) =$$

$$\left(\Sigma h \cap V(z_0)\right) \cup \left(\Gamma^1_{h,\mathbf{z}} \cap V\left(\frac{\partial h}{\partial z_0} + jaz_0^{j-1}\right)\right).$$

But, we just showed that $\Gamma^1_{h,\mathbf{z}} \cap V\left(\frac{\partial h}{\partial z_0} + jaz_0^{j-1}\right)$ is zero-dimensional at the origin.

Thus, near the origin, $\Sigma(h + az_0^j) = \Sigma h \cap V(z_0)$ and, by Proposition 1.16, has dimension equal to $s - 1$.

Proof of iv): Using iii), we have

$$\Gamma^i_{h+az_0^j,\tilde{\mathbf{z}}} = V\left(\frac{\partial h}{\partial z_{i+1}}, \ldots, \frac{\partial h}{\partial z_n}, \frac{\partial h}{\partial z_0} + jaz_0^{j-1}\right) \Big/ \Sigma(h + az_0^j) =$$

$$\left((\Gamma^{i+1}_{h,\mathbf{z}} \cup R) \cap V\left(\frac{\partial h}{\partial z_0} + jaz_0^{j-1}\right)\right) \Big/ (\Sigma h \cap V(z_0)),$$

where the ideal defining the scheme R consists of the intersection of those primary components, $\mathfrak{q}$, of any primary decomposition of the ideal $\left\langle \frac{\partial h}{\partial z_{i+1}}, \ldots, \frac{\partial h}{\partial z_n} \right\rangle$ such that $|V(\mathfrak{q})| \subseteq |\Sigma h|$. Regardless of the primary decomposition, $|R| \subseteq |\Sigma h|$ and so $\left|R \cap V\left(\frac{\partial h}{\partial z_0} + jaz_0^{j-1}\right)\right| \subseteq |\Sigma h \cap V(z_0)|$. Thus, by 1.2.ii, the previous equalities give us

$$\Gamma^i_{h+az_0^j,\tilde{\mathbf{z}}} = \left(\Gamma^{i+1}_{h,\mathbf{z}} \cap V\left(\frac{\partial h}{\partial z_0} + jaz_0^{j-1}\right)\right) \Big/ (\Sigma h \cap V(z_0)),$$

We would have our desired equality of cycles if we could show that no component of $\Gamma^{i+1}_{h,\mathbf{z}} \cap V\left(\frac{\partial h}{\partial z_0} + jaz_0^{j-1}\right)$ is contained in $\Sigma h \cap V(z_0)$.

But, the dimension of every component of $\Gamma^{i+1}_{h,\mathbf{z}} \cap V\left(\frac{\partial h}{\partial z_0} + jaz_0^{j-1}\right)$ is at least i, while we claim that the dimension at the origin of

$$\Gamma^{i+1}_{h,\mathbf{z}} \cap V\left(\frac{\partial h}{\partial z_0} + jaz_0^{j-1}\right) \cap \Sigma h \cap V(z_0)$$

is at most $i - 1$; for

$$\Gamma^{i+1}_{h,\mathbf{z}} \cap V\left(\frac{\partial h}{\partial z_0} + jaz_0^{j-1}\right) \cap \Sigma h \cap V(z_0) = \Gamma^{i+1}_{h,\mathbf{z}} \cap \Sigma h \cap V(z_0)$$

and, by Proposition 1.15, this equals $V(z_0) \cap \bigcup_{k \leqslant i} \Lambda^k_{h,\mathbf{z}}$. As all the Lê numbers of h are defined at the origin and $i \geqslant 1$, the dimension of $V(z_0) \cap \bigcup_{k \leqslant i} \Lambda^k_{h,\mathbf{z}}$ at the origin must at most $i - 1$.

Proof of v): Let $L = w - z_0$ and use $(L, z_0, \ldots, z_n)$ as coordinates for $\mathbb{C}^{n+2}$. Then,

$$\Gamma^1_{h+aw^j, w-z_0} = \Gamma^1_{h+a(L+z_0)^j, L} =$$

$$V\left(\frac{\partial h}{\partial z_0} + ja(L + z_0)^{j-1}, \frac{\partial h}{\partial z_1}, \ldots, \frac{\partial h}{\partial z_n}\right) \Big/ \Sigma(h + a(L + z_0)^j)$$

which, back in $(w, z_0, \dots, z_n)$ coordinates,

$$= V\left(\frac{\partial h}{\partial z_0} + jaw^{j-1}, \frac{\partial h}{\partial z_1}, \dots, \frac{\partial h}{\partial z_n}\right) \Big/ \Sigma(h + aw^j),$$

where $\Sigma(h + aw^j) = 0 \times \Sigma h$ as $j \geqslant 2$.

Hence, as schemes,

$$\Gamma^1_{h+aw^j, w-z_0} = \left(V\left(\frac{\partial h}{\partial z_0} + jaw^{j-1}\right) \cap \left(\mathbb{C} \times (\Gamma^1_{h,\mathbf{z}} \cup R)\right)\right) \Big/ (0 \times \Sigma h),$$

where, as sets, $R = \Sigma h$. Thus, since

$$\left|(\mathbb{C} \times R) \cap V\left(\frac{\partial h}{\partial z_0} + jaw^{j-1}\right)\right| \subseteq 0 \times \Sigma h,$$

we may use 1.2.ii to conclude that

$$\Gamma^1_{h+aw^j, w-z_0} = \left(V\left(\frac{\partial h}{\partial z_0} + jaw^{j-1}\right) \cap \left(\mathbb{C} \times \Gamma^1_{h,\mathbf{z}}\right)\right) \Big/ (0 \times \Sigma h).$$

But, no component of $V\left(\frac{\partial h}{\partial z_0} + jaw^{j-1}\right) \cap \left(\mathbb{C} \times \Gamma^1_{h,\mathbf{z}}\right)$ is contained in $0 \times \Sigma h$, for $V\left(\frac{\partial h}{\partial z_0} + jaw^{j-1}\right) \cap \left(\mathbb{C} \times \Gamma^1_{h,\mathbf{z}}\right)$ is purely one-dimensional at the origin while

$$\dim_{\mathbf{0}} V\left(\frac{\partial h}{\partial z_0} + jaw^{j-1}\right) \cap \left(\mathbb{C} \times \Gamma^1_{h,\mathbf{z}}\right) \cap (0 \times \Sigma h) =$$

$$\dim_{\mathbf{0}}\left(0 \times \left(V\left(\frac{\partial h}{\partial z_0}\right) \cap \Gamma^1_{h,\mathbf{z}} \cap \Sigma h\right)\right) = 0.$$

Therefore, as 1-cycles at the origin,

$$\Gamma^1_{h+aw^j, w-z_0} = V\left(\frac{\partial h}{\partial z_0} + jaw^{j-1}\right) \cdot \left(\mathbb{C} \times \Gamma^1_{h,\mathbf{z}}\right).$$

Now, we will apply the equality in 1.23 to split the calculation into two pieces. Using both the coordinates $(w, z_0, \dots, z_n)$ and $(L, z_0, \dots, z_n)$, we have:

$$\left(\Gamma^1_{h+aw^j, w-z_0} \cdot V(h + aw^j)\right)_{\mathbf{0}} = \left(\Gamma^1_{h+a(L+z_0)^j, L} \cdot V(h + a(L + z_0)^j)\right)_{\mathbf{0}} =$$

$$\left(\Gamma^1_{h+aw^j, w-z_0} \cdot V(L)\right)_{\mathbf{0}} + \left(\Gamma^1_{h+aw^j, w-z_0} \cdot V\left(\frac{\partial(h + a(L + z_0)^j)}{\partial L}\right)\right)_{\mathbf{0}} =$$

$$\left(V\left(\frac{\partial h}{\partial z_0} + jaw^{j-1}\right) \cdot \left(\mathbb{C} \times \Gamma^1_{h,\mathbf{z}}\right) \cdot V(w - z_0)\right)_{\mathbf{0}} +$$

$$\left(V\left(\frac{\partial h}{\partial z_0}+jaw^{j-1}\right)\cdot(\mathbb{C}\times\Gamma^1_{h,\mathbf{z}})\cdot V(jaw^{j-1})\right)_{\mathbf{0}}=$$

$$\left(\left(\mathbb{C}\times V\left(\frac{\partial h}{\partial z_0}+jaz_0^{j-1}\right)\right)\cdot(\mathbb{C}\times\Gamma^1_{h,\mathbf{z}})\cdot V(w-z_0)\right)_{\mathbf{0}}+$$

$$(j-1)\left(V\left(\frac{\partial h}{\partial z_0}+jaw^{j-1}\right)\cdot(\mathbb{C}\times\Gamma^1_{h,\mathbf{z}})\cdot V(w)\right)_{\mathbf{0}}=$$

$$\left(\left(\mathbb{C}\times\left(\Gamma^1_{h,\mathbf{z}}\cdot V\left(\frac{\partial h}{\partial z_0}+jaz_0^{j-1}\right)\right)\right)\cdot V(w-z_0)\right)_{\mathbf{0}}+$$

$$(j-1)\left(0\times\left(\Gamma^1_{h,\mathbf{z}}\cdot V\left(\frac{\partial h}{\partial z_0}\right)\right)\right)_{\mathbf{0}}.$$

However, by ii), $\Gamma^1_{h,\mathbf{z}}\cdot V\left(\frac{\partial h}{\partial z_0}+jaz_0^{j-1}\right)=\lambda^0_{h,\mathbf{z}}(\mathbf{0})[\mathbf{0}]$, and so

$$\left(\Gamma^1_{h+aw^j,w-z_0}\cdot V(h+aw^j)\right)_{\mathbf{0}}=\lambda^0_{h,\mathbf{z}}(\mathbf{0})+(j-1)\lambda^0_{h,\mathbf{z}}(\mathbf{0})=j\lambda^0_{h,\mathbf{z}}(\mathbf{0}).\ \square$$

Our next result will be to obtain the *generalized Lê-Iomdine formulas*; these formulas are a stunningly useful tool for reducing questions on general hypersurface singularities to the much easier case of isolated hypersurface singularities. The formulas tell how the Lê numbers of h change when a large power of one of the variables is added. By 4.3.iii, this modification of the function h will have a critical locus of dimension one smaller than that of h itself. Proceeding inductively, one arrives at the case of an isolated singularity.

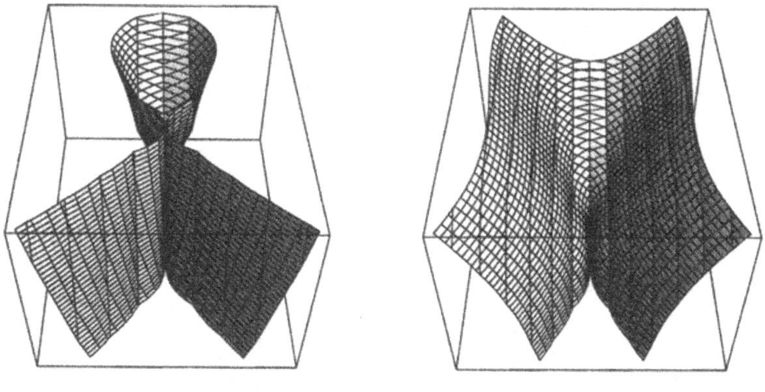

Figure 4.4. The effect of adding a large power of a variable

Theorem 4.5. *Let $j \geqslant 2$, let $h : (\mathcal{U}, \mathbf{0}) \to (\mathbb{C}, 0)$ be an analytic function, let s denote $\dim_{\mathbf{0}} \Sigma h$, and assume that $s \geqslant 1$. Let $\mathbf{z} = (z_0, \ldots, z_n)$ be a linear choice of coordinates such that $\lambda^i_{h,\mathbf{z}}(\mathbf{0})$ is defined for all $i \leqslant s$. Let a be a non-zero complex number, and use the coordinates $\tilde{\mathbf{z}} = (z_1, \ldots, z_n, z_0)$ for $h + az_0^j$.*

If j is greater than or equal to the maximum polar ratio for h then, for all but a finite number of complex a, $\Sigma(h + az_0^j) = \Sigma h \cap V(z_0)$ as germs of sets at $\mathbf{0}$, $\dim_{\mathbf{0}} \Sigma(h + az_0^j) = s - 1$, $\lambda^i_{h+az_0^j,\tilde{\mathbf{z}}}(\mathbf{0})$ exists for all $i \leqslant s - 1$, and

$$\lambda^0_{h+az_0^j,\tilde{\mathbf{z}}}(\mathbf{0}) = \lambda^0_{h,\mathbf{z}}(\mathbf{0}) + (j-1)\lambda^1_{h,\mathbf{z}}(\mathbf{0}),$$

and, for $1 \leqslant i \leqslant s - 1$,

$$\lambda^i_{h+az_0^j,\tilde{\mathbf{z}}}(\mathbf{0}) = (j-1)\lambda^{i+1}_{h,\mathbf{z}}(\mathbf{0}).$$

Moreover, if we have the strict inequality that j is greater than the maximum polar ratio for h, then the above equalities hold for all non-zero a; in particular, this is the case if $j \geqslant 2 + \lambda^0_{h,\mathbf{z}}(\mathbf{0})$.

Proof. Despite the fact that we have improved the statement on the bounds, the proof is exactly that of Theorem 3.3 of [**Mas5**]. We will use the results of the lemma without further comment.

$$\lambda^0_{h+az_0^j,\tilde{\mathbf{z}}}(\mathbf{0}) = \left(\Gamma^1_{h+az_0^j,\tilde{\mathbf{z}}} \cdot V\left(\frac{\partial h}{\partial z_1}\right)\right)_{\mathbf{0}} =$$

$$\left(\Gamma^2_{h,\mathbf{z}} \cdot V\left(\frac{\partial h}{\partial z_0} + jaz_0^{j-1}\right) \cdot V\left(\frac{\partial h}{\partial z_1}\right)\right)_{\mathbf{0}} =$$

$$\left(\left(\Gamma^1_{h,\mathbf{z}} + \Lambda^1_{h,\mathbf{z}}\right) \cdot V\left(\frac{\partial h}{\partial z_0} + jaz_0^{j-1}\right)\right)_{\mathbf{0}} =$$

$$\left(\Gamma^1_{h,\mathbf{z}} \cdot V\left(\frac{\partial h}{\partial z_0} + jaz_0^{j-1}\right)\right)_{\mathbf{0}} + (j-1)\left(\Lambda^1_{h,\mathbf{z}} \cdot V(z_0)\right)_{\mathbf{0}} =$$

$$\lambda^0_{h,\mathbf{z}}(\mathbf{0}) + (j-1)\lambda^1_{h,\mathbf{z}}(\mathbf{0}).$$

For $i \geqslant 1$,

$$\Gamma^i_{h+az_0^j,\tilde{\mathbf{z}}} = \Gamma^{i+1}_{h,\mathbf{z}} \cdot V\left(\frac{\partial h}{\partial z_0} + jaz_0^{j-1}\right)$$

and so

$$\Gamma^i_{h+az_0^j,\tilde{\mathbf{z}}} + \Lambda^i_{h+az_0^j,\tilde{\mathbf{z}}} = \Gamma^{i+1}_{h+az_0^j,\tilde{\mathbf{z}}} \cdot V\left(\frac{\partial h}{\partial z_{i+1}}\right) =$$

$$\Gamma^{i+2}_{h,\mathbf{z}} \cdot V\left(\frac{\partial h}{\partial z_0} + jaz_0^{j-1}\right) \cdot V\left(\frac{\partial h}{\partial z_{i+1}}\right) =$$

$$\left(\Gamma^{i+1}_{h,\mathbf{z}} + \Lambda^{i+1}_{h,\mathbf{z}}\right) \cdot V\left(\frac{\partial h}{\partial z_0} + jaz_0^{j-1}\right) =$$

$$\Gamma^i_{h+az_0^j,\tilde{\mathbf{z}}} + (j-1)\left(\Lambda^{i+1}_{h,\mathbf{z}} \cdot V(z_0)\right).$$

Cancelling $\Gamma^i_{h+az_0^j,\tilde{\mathbf{z}}}$ from each side of the above equation, we find

$$\Lambda^i_{h+az_0^j,\tilde{\mathbf{z}}} = (j-1)\left(\Lambda^{i+1}_{h,\mathbf{z}} \cdot V(z_0)\right).$$

Therefore, for $i \geqslant 1$, $\lambda^i_{h+az_0^j,\tilde{\mathbf{z}}}(\mathbf{0}) = (j-1)\lambda^{i+1}_{h,\mathbf{z}}(\mathbf{0})$. □

By applying the Lê-Iomdine formulas inductively, we immediately conclude

Corollary 4.6. *Let $h : (\mathcal{U}, \mathbf{0}) \to (\mathbb{C}, 0)$ be an analytic function, let s denote $\dim_{\mathbf{0}}\Sigma h$, and let $\mathbf{z} = (z_0, \dots, z_n)$ be a linear choice of coordinates such that $\lambda^i_{h,\mathbf{z}}(\mathbf{0})$ is defined for all $i \leqslant s$. Then, for $0 \ll j_0 \ll j_1 \ll \dots \ll j_{s-1}$,*

$$h + z_0^{j_0} + z_1^{j_1} + \dots + z_{s-1}^{j_{s-1}}$$

has an isolated singularity at the origin, and its Milnor number is given by

$$\mu(h + z_0^{j_0} + z_1^{j_1} + \dots + z_{s-1}^{j_{s-1}}) = \sum_{i=0}^{s}\left(\lambda^i_{h,\mathbf{z}}(\mathbf{0})\prod_{k=0}^{i-1}(j_k - 1)\right) =$$

$$\lambda^0_{h,\mathbf{z}}(\mathbf{0}) + (j_0-1)\lambda^1_{h,\mathbf{z}}(\mathbf{0}) + (j_1-1)(j_0-1)\lambda^2_{h,\mathbf{z}}(\mathbf{0}) + \dots$$
$$+(j_{s-1}-1)\dots(j_1-1)(j_0-1)\lambda^s_{h,\mathbf{z}}(\mathbf{0}).$$

As another quick application of the Lê-Iomdine formulas, we have the following Plücker formula.

Corollary 4.7. *Let h be a homogeneous polynomial of degree d in $n+1$ variables, let $s = \dim_{\mathbf{0}}\Sigma h$, and suppose that $\lambda^i_{h,\mathbf{z}}(\mathbf{0})$ exists for all $i \leqslant s$. Then,*

$$\sum_{i=0}^{s}(d-1)^i\lambda^i_{h,\mathbf{z}}(\mathbf{0}) = (d-1)^{n+1}.$$

Proof. By Remark 4.2, the maximum polar ratio is d. By an inductive application of the Lê-Iomdine formulas, we arrive at a function,

$$f := h + a_0 z_0^d + a_1 z_1^d + \dots + a_{s-1}z_{s-1}^d,$$

with an isolated singularity at the origin and such that the Milnor number of $f = \lambda^0_f(\mathbf{0}) = \sum_{i=0}^{s}(d-1)^i\lambda^i_{h,\mathbf{z}}(\mathbf{0})$. Now, by [**M-O**], this Milnor number is precisely $(d-1)^{n+1}$. □

In Chapter 9, we will see that the Lê cycles are actually Segre cycles. Knowing this, the above Plücker formula is a special case of a much more general result of Van Gastel [**Gas1**, 1.2.c].

Remark 4.8. In general, Proposition 4.7 makes it slightly easier to calculate the Euler characteristic of the Milnor fibre of a homogeneous polynomial. In the case of a one-dimensional critical locus, 4.7 tells us that, if we know the degree and λ^1, then we know λ^0 and, hence, the Euler characteristic (see also [**M-S**] and [**Si2**]).

Being able to calculate the Euler characteristic of the Milnor fibre of a homogeneous singularity implies in many cases that we can also calculate the Euler characteristic of the Milnor fibre of a weighted-homogeneous polynomial.

To see this, let $f : \mathbb{C}^{n+1} \to \mathbb{C}$ be a weighted homogeneous polynomial. Then, there exist positive integers $r_0, \ldots, r_n$ such that, if $\pi : \mathbb{C}^{n+1} \to \mathbb{C}^{n+1}$ is given by

$$\pi(z_0, \ldots, z_n) = (z_0^{r_0}, \ldots, z_n^{r_n}),$$

then $h := f \circ \pi$ is homogeneous. Moreover, the restriction of π induces a map from the Milnor fibre, $F_{h,\mathbf{0}}$, of h at the origin to the Milnor fibre, $F_{f,\mathbf{0}}$, of f at the origin; denote this map by $\tilde{\pi} : F_{h,\mathbf{0}} \to F_{f,\mathbf{0}}$.

Now, consider the stratification of $\mathbb{C}^{n+1}$ derived from the hyperplane arrangement given by all the coordinate hyperplanes. That is, let I denote the indexing set $\{0, \ldots, n\}$, and for each $J \subseteq I$, let w_J denote the intersection of hyperplanes (a.k.a. the *flat*) given by

$$w_J := V(z_j \mid j \in J),$$

and let S_J denote the Whitney stratum

$$S_J := w_J - \bigcup_{J \subsetneq K} w_K.$$

The stratification $\{S_J\}$ determines Whitney stratifications $\{S_J \cap F_{h,\mathbf{0}}\}$ and $\{S_J \cap F_{f,\mathbf{0}}\}$ of $F_{h,\mathbf{0}}$ and $F_{f,\mathbf{0}}$, respectively, and with these stratifications, $\tilde{\pi}$ becomes a stratified map. Moreover, the restriction of $\tilde{\pi}$ to a map from $S_J \cap F_{h,\mathbf{0}}$ to $S_J \cap F_{f,\mathbf{0}}$ is a topological covering map with fibre equal to $\prod_{i \notin J} r_i$ points.

Hence,

$$\chi(F_{h,\mathbf{0}}) = \sum_J \chi(S_J \cap F_{h,\mathbf{0}}) = \sum_J \left(\prod_{i \notin J} r_i\right) \chi(S_J \cap F_{f,\mathbf{0}}).$$

Some elementary combinatorics shows that this last quantity is equal to

$$\sum_{J} c_{J}\left(\sum_{J\subseteq K}\chi(S_{K}\cap F_{f,\mathbf{0}})\right),$$

where

$$c_{J} := (-1)^{|J|}\sum_{L\subseteq J}\left((-1)^{|L|}\prod_{i\notin L} r_i\right).$$

The advantage of this last form is that

$$\sum_{J\subseteq K}\chi(S_{K}\cap F_{f,\mathbf{0}}) = \chi(F_{f_{|w_J},\mathbf{0}}).$$

Therefore, we have that

$$\chi(F_{h,\mathbf{0}}) = \sum_{J} c_{J}\chi(F_{f_{|w_J},\mathbf{0}}),$$

where c_J is as above.

It follows that

$$\chi(F_{h,\mathbf{0}}) = (r_0\dots r_n)\chi(F_{f,\mathbf{0}}) + \sum_{J\neq\emptyset} c_{J}\chi(F_{f_{|w_J},\mathbf{0}}),$$

and so, finally, we arrive at the formula

$$\chi(F_{f,\mathbf{0}}) = \frac{\chi(F_{h,\mathbf{0}}) - \sum_{J\neq\emptyset} c_{J}\chi(F_{f_{|w_J},\mathbf{0}})}{r_0\dots r_n}.$$

This formula is inductively useful since, if $J\neq\emptyset$, then $f_{|w_J}$ is a weighted-homogeneous polynomial in fewer variables (compare with [**Di**]). Note that, in this formula, we need not consider the term where $J=\{0,\dots,n\}$, for then $w_J = \mathbf{0}$ and hence $\chi(F_{f_{|w_J},\mathbf{0}}) = 0$.

This is particularly useful in the case where f is a weighted-homogeneous polynomial with a one-dimensional critical locus and each restriction to a flat, $f_{|w_J}$, also has a one-dimensional (or zero-dimensional) critical locus.

Example 4.9. For instance, consider the case of a possibly non-reduced, weighted-homogeneous plane curve singularity. Suppose that the irreducible factorization of $f(z_0,z_1)$ is $z_0^a z_1^b\prod f_i^{m_i}$, where we allow for the case where a or b equals 0. Let $\pi(z_0,z_1) = (z_0^{r_0}, z_1^{r_1})$, and let h denote the homogeneous polynomial $f\circ\pi$. Let h_i denote the homogeneous polynomial $f_i\circ\pi$. Let d be the degree of h and let d_i be the degree of h_i.

Then, the formula of 4.7 becomes

$$\chi(F_{f,\mathbf{0}}) = \frac{\chi(F_{h,\mathbf{0}}) + (r_0-1)r_1\chi(F_{f_{|V(z_0)},\mathbf{0}}) + (r_1-1)r_0\chi(F_{f_{|V(z_1)},\mathbf{0}})}{r_0r_1}.$$

Now, $\chi(F_{f_{|V(z_k)},\mathbf{0}}) = 0$ if $f_{|V(z_k)} \equiv 0$ and simply equals the multiplicity of $f_{|V(z_k)}$ otherwise. In addition, as h is homogeneous, we may calculate $\chi(F_{h,\mathbf{0}})$ from 4.6 by knowing only $\lambda^1_h(\mathbf{0})$, which we may calculate as in 2.9.

We find easily that

$$r_0r_1\chi(F_{f,\mathbf{0}}) = \begin{cases} -d\sum d_i, & \text{if } a \neq 0, b \neq 0 \\ d(r_0 - \sum d_i), & \text{if } a = 0, b \neq 0 \\ d(r_1 - \sum d_i), & \text{if } a \neq 0, b = 0 \\ d(r_0 + r_1 - \sum d_i), & \text{if } a = b = 0. \end{cases}$$

Example 4.10. We can also apply the formula of 4.8 in harder cases. Consider the swallowtail singularity; this is given as the zero locus of

$$f = 256z_0^3 - 27z_1^4 - 128z_0^2z_2^2 + 144z_0z_1^2z_2 + 16z_0z_2^4 - 4z_1^2z_2^3$$

(see, for instance, [**Te1**]).

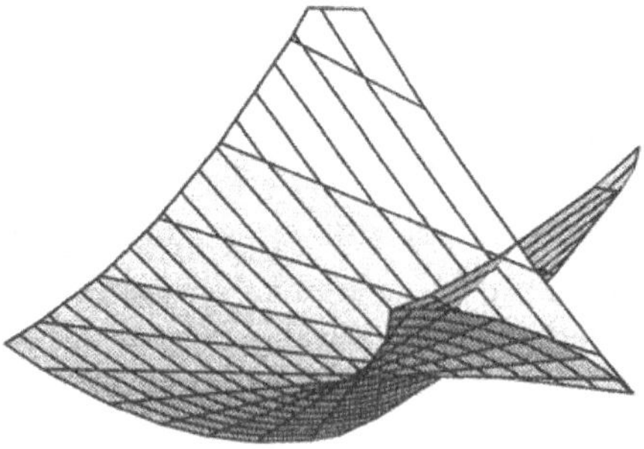

Figure 4.11. The swallowtail singularity

If $\pi(z_0, z_1, z_2) = (z_0^4, z_1^3, z_2^2)$, then $h = f \circ \pi$ is homogeneous of degree 12. Using the notation of 4.8, we find

$$c_{\{0\}} = (-1)(4 \cdot 3 \cdot 2 - 3 \cdot 2) = -18.$$

$$f_{|_{w_{\{0\}}}} = -27z_1^4 - 4z_1^2z_2^3 = z_1^2(-27z_1^2 - 4z_2^3).$$

Hence, from 4.9, we find that

$$\chi(F_{f_{|_{w_{\{0\}}}},\mathbf{0}}) = -8.$$

Similarly,

$$c_{\{1\}} = (-1)(4\cdot 3\cdot 2 - 4\cdot 2) = -16.$$

$$f_{|_{w_{\{1\}}}} = 256z_0^3 - 128z_0^2z_2^2 + 16z_0z_2^4 = 16z_0(16z_0^2 - 8z_0z_2^2 + z_2^4) =$$

$$16z_0(4z_0 - z_2^2)^2.$$

$$\chi(F_{f_{|_{w_{\{1\}}}},\mathbf{0}}) = -3.$$

and

$$c_{\{2\}} = (-1)(4\cdot 3\cdot 2 - 4\cdot 3) = -12.$$

$$f_{|_{w_{\{2\}}}} = 256z_0^3 - 27z_1^4.$$

$$\chi(F_{f_{|_{w_{\{2\}}}},\mathbf{0}}) = -5.$$

We also find

$$f_{|_{w_{\{0,1\}}}} \equiv 0,$$

$$c_{\{0,2\}} = 4\cdot 3\cdot 2 - 3\cdot 2 - 4\cdot 3 + 3 = 9,$$

$$f_{|_{w_{\{0,2\}}}} = -27z_1^4,$$

$$\chi(F_{f_{|_{w_{\{0,2\}}}},\mathbf{0}}) = 4,$$

and

$$c_{\{1,2\}} = 4\cdot 3\cdot 2 - 4\cdot 2 - 4\cdot 3 + 4 = 8,$$

$$f_{|_{w_{\{1,2\}}}} = 256z_0^3,$$

$$\chi(F_{f_{|_{w_{\{0,2\}}}},\mathbf{0}}) = 3.$$

Having made these calculations, it still remains for us to calculate $\chi(F_{h,\mathbf{0}})$. We can do this using 4.7, provided that we know that h has a one-dimensional critical locus and provided that we can calculate $\lambda_h^1(\mathbf{0})$. While this calculation can be made by hand, it is rather tedious; a computer algebra program – such as *Macaulay*, a public domain program written by Michael Stillman and Dave Bayer – can tell us that not only does h have a one-dimensional critical locus, but that the multiplicity of the Jacobian scheme at the origin is 83, i.e. $\lambda^1_h(\mathbf{0}) = 83$. Therefore,

$$\chi(F_{h,\mathbf{0}}) = \lambda^0_h(\mathbf{0}) - \lambda^1_h(\mathbf{0}) + 1 = d\lambda^1_h(\mathbf{0}) - (d-1)^{n+1} + 1 =$$

$$12 \cdot 83 - 11^3 + 1 = 336.$$

Finally,

$$\chi(F_{f,\mathbf{0}}) = \frac{336 - (-18)(-8) - (-16)(-3) - (-12)(-5) - 9 \cdot 4 - 8 \cdot 3}{4 \cdot 3 \cdot 2} = 1.$$

Remark 4.12. In the above example, we resorted to a computer calculation at one point. If we are willing to use a computer algebra program at each step, then there is a much easier way to calculate the Euler characteristic of the Milnor fibre in the case of a one-dimensional critical locus – whether the function, h, is a weighted-homogeneous polynomial or not. This is the method that we describe in [**M-S**].

Any computer program which can calculate the multiplicities of ideals in a polynomial ring, given a set of generators, can calculate the Lê numbers of a polynomial. (A number of programs have this capability, but by far the most efficient that we know of is *Macaulay*.)

Given such a program and a polynomial, h, with a one-dimensional singular set, one proceeds as follows to calculate the Lê numbers, λ^0 and λ^1, at the origin with respect to a generic set of coordinates.

As we saw in 1.19, λ^1 is nothing other than the multiplicity of the Jacobian scheme of h at the origin. So, one can have the program calculate it.

Now, we need a hyperplane that is generic enough so that its intersection number (at the origin) with the (reduced) singular set is, in fact, equal to the multiplicity of the singular set. Usually, one knows the singular set (as a set) well enough to know such a hyperplane. (Alternatively, there are programs which can find the singular set for you – though how they present the answer is not always helpful.) We shall assume now, in addition to having λ^1, that we also have such a hyperplane, $V(L)$, for some linear form, L .

By the work of Iomdine [**Io**] and Lê [**Lê2**] (or our generalization in 4.5 or [**M-S**]), we have that: for all k sufficiently large, $h + L^k$ has an isolated singularity at the origin and the Milnor number $\mu(h + L^k)$ equals $\lambda^0 + (k-1)\lambda^1$. But, the Milnor number is again nothing other than the multiplicity of the Jacobian scheme at the origin, and so we may use our program to calculate it. Thus, we can find λ^0 – provided that we have an effective method for knowing when we have chosen k large enough so that the formula of Iomdine and Lê holds.

However, we have such a method. If $h + L^k$ has an isolated singularity, let μ_k denote its Milnor number. (Given a particular k, one must either check by hand whether $h + L^k$ has an isolated singularity or have a program do it. *Macaulay* will tell you the dimension of the singular set in the course of calculating the multiplicity of the Jacobian scheme at the origin.) A quick look at the proof of the Iomdine-Lê formula in 4.5 shows that the formula holds provided that

$$\mu_k - (k-1)\lambda^1 \leqslant k-2.$$

Therefore, to find λ^0, one starts with a relatively small k and checks whether $\mu_k \leqslant k - 2 + (k-1)\lambda^1$. If the inequality is false, pick a larger k. Eventually, the inequality will hold and then

$$\lambda^0 = \mu_k - (k-1)\lambda^1.$$

There is a similar alternative method for calculating not only the Lê numbers but also the maximum polar ratio of h. Again, from the proof of 4.5, it is not difficult to see that, if μ_i is as above, then $\mu_{k+1} - \mu_k = \lambda^1$ if and only if $k \geqslant$ the maximum polar ratio.

Hence, to find the maximum polar ratio, one calculates μ_k for successive values of k – looking for a difference of λ^1. Once this occurs, $k \geqslant$ the maximum polar ratio and, as before, we conclude that

$$\lambda^0 = \mu_k - (k-1)\lambda^1.$$

Note, moreover, that this method works whether $\lambda^1 =$ the multiplicity of the Jacobian scheme at the origin or not. As $\lambda^1 \geqslant$ the multiplicity of the Jacobian scheme at the origin, with equality being the generic case, it follows that if $\mu_{k+1} - \mu_k =$ the multiplicity of the Jacobian scheme at the origin, then $\lambda^1 =$ the multiplicity of the Jacobian scheme at the origin and $\lambda^0 = \mu_k - (k-1)\lambda^1$.

While this method requires one to calculate at least two Milnor numbers, μ_k, it will still be a more efficient way of calculating λ^0 – provided that the maximum polar ratio is significantly smaller than λ^0 itself. This would be the case, for instance, if the polar curve had a large number of components.

Consider again the swallowtail of Example 4.10 defined by

$$h = 256z_0^3 - 27z_1^4 - 128z_0^2z_2^2 + 144z_0z_1^2z_2 + 16z_0z_2^4 - 4z_1^2z_2^3.$$

We use *Macaulay* to find that the multiplicity of the Jacobian scheme at the origin equals 5. (Alternatively, we know that the singularities of the swallowtail consist of a smooth curve of ordinary double points plus a multiplicity two curve of cusps; hence, the multiplicity of the Jacobian scheme at the origin $= 1 + 2 \cdot 2 = 5$.) Now, using the notation above and letting $L = z_2$, we find

$k =$	2	3	4	5	6	7
$\mu_k =$	6	12	18	24	30	35

From the table, we see that the maximum polar ratio is at most 6, λ^1 is, in fact, equal to 5, and $\lambda^0 = 30 - (6-1)5 = 5$. Hence, the Euler characteristic of the Milnor fibre of h equals $\lambda^0 - \lambda^1 + 1 = 1$, which agrees with our previous calculation.

As the swallowtail is such an important singularity, one might wonder how close the Morse inequalities of Theorem 3.3 are to being equalities in this example. The answer is: not very. A. Suciu informs us that the degree 1 and 2 homology groups of the Milnor fibre of the swallowtail at the origin are both free Abelian of rank 2.

Why are the Lê numbers off by so much from the Betti numbers? It is because the Lê numbers record information that the Betti numbers do not.

In the case of the swallowtail, λ^1 records the information that there is an entire cusp of cusp singularities coming into the origin plus a line of quadratic singularities. Thus, λ^1 equals (the multiplicity of the cusp)(the Milnor number of the cusp) plus (the Milnor number of the quadratic singularity) = $(2)(2)+1=5$. Now, as λ^1 is forced to be 5, λ^0 also has to be 5 - in order to make $\lambda^0 - \lambda^1 + 1$ come out to equal the Euler characteristic.

Remark 4.13. For a projective hypersurface, X, defined by a homogeneous polynomial, h, in the projective coordinates $(z_0 : \cdots : z_n)$, one may ask if there is some reasonable notion of **global** Lê numbers.

It is easy to see that if one takes affine patches on X, calculates λ^0 at points of each patch with respect to generic coordinates, and adds together the finite number of non-zero results that one gets, then the answer is precisely $\lambda^1_h(\mathbf{0})$ (in the ordinary, affine sense) with respect to generic coordinates. It seems reasonable, then, to define the global λ^i_X to be $\lambda^{i+1}_h(\mathbf{0})$.

If one makes this definition, it might initially look as though $\lambda^0_h(\mathbf{0})$ should provide a new interesting invariant of X. However, Corollary 4.7 tells us that $\lambda^0_h(\mathbf{0})$ can be calculated from the higher Lê numbers together with the degree of X.

We shall now prove a uniform version of the generalized Lê-Iomdine formulas for one-parameter families of germs of hypersurface singularities at the origin, i.e. $\overset{\circ}{\mathbb{D}}$ will be an open disc about the origin in $\mathbb{C}$, $\mathcal{U}$ will be an open neighborhood of the origin in $\mathbb{C}^{n+1}$ and $f : (\overset{\circ}{\mathbb{D}} \times \mathcal{U}, \overset{\circ}{\mathbb{D}} \times \mathbf{0}) \to (\mathbb{C}, 0)$ will be an analytic function; naturally, we write f_t for the function defined by $f_t(\mathbf{z}) := f(t, \mathbf{z})$.

First, we need a lemma

Lemma 4.14. *For all i and for all $\mathbf{p} = (t_0, z_0, \ldots, z_n)$ near the origin such that $t_0 \neq 0$,*

$$\Gamma^i_{f_{t_0}, \mathbf{z}} = \Gamma^{i+1}_{f,(t,\mathbf{z})} \cap V(t - t_0) = \Gamma^{i+1}_{f,(t,\mathbf{z})} \cdot V(t - t_0)$$

as cycles at $\mathbf{p}$, *regardless of how generic* $(t, z_0, \ldots, z_n)$ *may be.*

Proof. Fix any good stratification, $\mathfrak{S}$, for f at the origin. The stratified critical values of the function t are isolated; hence, in a neighborhood of the origin, the map t restricted to each of the strata of $\mathfrak{S}$ can have only 0 as a critical value (see [**Mas5**, Prop. 1.3]). Therefore, for all small $t_0 \neq 0$, $V(t - t_0)$ is a prepolar slice of f at $\mathbf{p}$. In particular, $\Sigma f \cap V(t - t_0) = \Sigma(f_{|V(t-t_0)})$.

Thus,

$$\Gamma^i_{f_{t_0},\mathbf{z}} = V\left(t - t_0, \frac{\partial f}{\partial z_i}, \ldots, \frac{\partial f}{\partial z_n}\right) \Big/ \Sigma(f_{t_0}) =$$

$$V\left(t - t_0, \frac{\partial f}{\partial z_i}, \ldots, \frac{\partial f}{\partial z_n}\right) \Big/ \Sigma f \cap V(t - t_0) =$$

$$V\left(t - t_0, \frac{\partial f}{\partial z_i}, \ldots, \frac{\partial f}{\partial z_n}\right) \Big/ \Sigma f = \left(\Gamma^{i+1}_{f,(t,\mathbf{z})} \cap V(t - t_0)\right) \Big/ \Sigma f,$$

where the last equality uses 1.2.i.

But, we claim that this equals $\Gamma^{i+1}_{f,(t,\mathbf{z})} \cap V(t - t_0)$ up to embedded subvariety for small $t_0 \neq 0$. For otherwise, $\Gamma^{i+1}_{f,(t,\mathbf{z})} \cap V(t - t_0)$ would have a component contained in Σf for an infinite number of small $t_0 \neq 0$, which would imply that $\Gamma^{i+1}_{f,(t,\mathbf{z})}$ has a component contained in Σf - a contradiction of the definition of $\Gamma^{i+1}_{f,(t,\mathbf{z})}$. Therefore, $\Gamma^i_{f_{t_0},\mathbf{z}} = \Gamma^{i+1}_{f,(t,\mathbf{z})} \cap V(t - t_0)$ up to embedded subvariety, and the conclusion follows easily. □

As in Theorem 4.5, when we use the coordinates $\mathbf{z} = (z_0, \ldots, z_n)$ for f_t, we use the rotated coordinates $\tilde{\mathbf{z}} = (z_1, z_2, \ldots, z_n, z_0)$ for $f_t + z_0^j$.

Theorem 4.15. (Uniform Lê-Iomdine formulas) *Let* $s := \dim_{\mathbf{0}} \Sigma f_0$, *and suppose that* $s \geqslant 1$. *Suppose that* $\lambda^i_{f_t,\mathbf{z}}(\mathbf{0})$ *is defined for all* $i \leqslant s$ *and for all small* t. *Then, there exist* $\tau > 0$ *and* j_0 *such that, for all* $j \geqslant j_0$ *and for all* $t \in \mathring{\mathbb{D}}_\tau$, $\dim_{\mathbf{0}} \Sigma(f_0 + z_0^j) = s - 1$, $\lambda^i_{f_t + z_0^j, \tilde{\mathbf{z}}}(\mathbf{0})$ *is defined for all* $i \leqslant s - 1$, *and*

i) $\quad \lambda^0_{f_t + z_0^j, \tilde{\mathbf{z}}}(\mathbf{0}) = \lambda^0_{f_t,\mathbf{z}}(\mathbf{0}) + (j - 1)\lambda^1_{f_t,\mathbf{z}}(\mathbf{0})$;

ii) $\quad \lambda^i_{f_t + z_0^j, \tilde{\mathbf{z}}}(\mathbf{0}) = (j - 1)\lambda^{i+1}_{f_t,\mathbf{z}}(\mathbf{0})$, *for* $1 \leqslant i \leqslant s - 1$;

iii) $\quad \Sigma(f_t + z_0^j) = \Sigma f_t \cap V(z_0)$ *near* $\mathbf{0}$.

Proof. Given 4.5, all that we must show is that $\{\lambda^0_{f_{t_0},\mathbf{z}}(\mathbf{0})\}_{t_0}$ is bounded for small t_0. Clearly, it suffices to show that $\{\lambda^0_{f_{t_0},\mathbf{z}}(\mathbf{0})\}_{t_0}$ is bounded for small $t_0 \neq 0$. Of course, what we actually show is that, for small $t_0 \neq 0$, $\lambda^0_{f_{t_0},\mathbf{z}}(\mathbf{0})$ is independent of t_0.

For small $t_0 \neq 0$, we may apply the lemma to conclude

$$\Lambda^0_{f_{t_0},\mathbf{z}} = \Gamma^1_{f_{t_0},\mathbf{z}} \cdot V\left(\frac{\partial f}{\partial z_0}\right) =$$

$$\Gamma^2_{f,(t,\mathbf{z})} \cdot V(t-t_0) \cdot V\left(\frac{\partial f}{\partial z_0}\right) = \left(\Gamma^1_{f,(t,\mathbf{z})} + \Lambda^1_{f,(t,\mathbf{z})}\right) \cdot V(t-t_0).$$

Thus, $\Gamma^1_{f,(t,\mathbf{z})} + \Lambda^1_{f,(t,\mathbf{z})}$ has a one-dimensional component, $n_\nu[\nu]$, which coincides with $\mathbb{C} \times \mathbf{0}$ near $\mathbf{0}$ (and so, must actually be a component of $\Lambda^1_{f,(t,\mathbf{z})}$) and such that $\lambda^0_{f_{t_0},\mathbf{z}}(\mathbf{0}) = \left(n_\nu[\nu] \cdot V(t-t_0)\right)_{(t_0,\mathbf{0})} = n_\nu$ for all small non-zero t_0. The conclusion follows. □

As we saw in Example 2.10, the Lê numbers in a family are not individually upper-semicontinuous. However, we do have the following.

Corollary 4.16. *Using the notation of the theorem, the tuple of Lê numbers*

$$\left(\lambda^s_{f_t,\mathbf{z}}(\mathbf{0}), \lambda^{s-1}_{f_t,\mathbf{z}}(\mathbf{0}), \dots, \lambda^0_{f_t,\mathbf{z}}(\mathbf{0})\right)$$

is lexigraphically upper-semicontinuous in the t variable, i.e. for all t small, either

$$\lambda^s_{f_0,\mathbf{z}}(\mathbf{0}) > \lambda^s_{f_t,\mathbf{z}}(\mathbf{0})$$

or

$$\lambda^s_{f_0,\mathbf{z}}(\mathbf{0}) = \lambda^s_{f_t,\mathbf{z}}(\mathbf{0}) \quad \textit{and} \quad \lambda^{s-1}_{f_0,\mathbf{z}}(\mathbf{0}) > \lambda^{s-1}_{f_t,\mathbf{z}}(\mathbf{0})$$

or

$$\vdots$$

or

$$\lambda^s_{f_0,\mathbf{z}}(\mathbf{0}) = \lambda^s_{f_t,\mathbf{z}}(\mathbf{0}),\ \lambda^{s-1}_{f_0,\mathbf{z}}(\mathbf{0}) = \lambda^{s-1}_{f_t,\mathbf{z}}(\mathbf{0}), \dots,\ \lambda^1_{f_0,\mathbf{z}}(\mathbf{0}) = \lambda^1_{f_t,\mathbf{z}}(\mathbf{0}),$$
$$\textit{and } \lambda^0_{f_0,\mathbf{z}}(\mathbf{0}) \geqslant \lambda^0_{f_t,\mathbf{z}}(\mathbf{0}).$$

Proof. By applying 4.15 inductively, as in 4.6, we find that, if $0 \ll j_0 \ll j_1 \ll \cdots \ll j_{s-1}$, then, for all small t, $f_t + z_0^{j_0} + z_1^{j_1} + \cdots + z_{s-1}^{j_{s-1}}$ has an isolated singularity at the origin, and its Milnor number is given by

$$\mu(f_t + z_0^{j_0} + z_1^{j_1} + \cdots + z_{s-1}^{j_{s-1}}) =$$

$$\lambda^0_{f_t,\mathbf{z}}(\mathbf{0}) + (j_0 - 1)\lambda^1_{f_t,\mathbf{z}}(\mathbf{0}) + (j_1 - 1)(j_0 - 1)\lambda^2_{f_t,\mathbf{z}}(\mathbf{0}) + \dots$$
$$+ (j_{s-1} - 1)\dots(j_1 - 1)(j_0 - 1)\lambda^s_{f_t,\mathbf{z}}(\mathbf{0}).$$

Now, as the Milnor number is upper-semicontinuous, the conclusion is immediate. □

Before we leave this chapter, we want to see how adding a large power of z_0 affects the prepolarity condition.

Proposition 4.17. *Let $\mathfrak{G}$ be a good stratification for h at $\mathbf{0}$ and let $V(z_0)$ be a prepolar slice with respect to $\mathfrak{G}$ at the origin. Suppose $a \neq 0$ and that j is such that $\Sigma(h + az_0^j) = \Sigma h \cap V(z_0)$ as sets. Then,*

$$\mathfrak{G}' = \{V(h + az_0^j) - \Sigma h \cap V(z_0)\} \cup \{G \cap V(z_0) \mid G \text{ is a singular stratum of } \mathfrak{G}\}$$

is a good stratification for $h + az_0^j$ at $\mathbf{0}$.

Proof. Suppose we have $\mathbf{p}_i \notin \Sigma h \cap V(z_0)$ such that $\mathbf{p}_i \to \mathbf{p} \in G \cap V(z_0)$, where G is a singular stratum, and such that $T_{\mathbf{p}_i} V(h + az_0^j - (h + az_0^j)_{|\mathbf{p}_i}) \to \mathcal{T}$. We wish to show that $T_{\mathbf{p}}(G \cap V(z_0)) = T_{\mathbf{p}} G \cap T_{\mathbf{p}} V(z_0) \subseteq \mathcal{T}$.

If $\mathcal{T} = T_{\mathbf{p}} V(z_0)$ of $G = \{\mathbf{0}\}$, then we are finished. So suppose that $\mathcal{T} \neq T_{\mathbf{p}} V(z_0)$and $G \neq \{\mathbf{0}\}$. Then, for all but a finite number of i, $T_{\mathbf{p}_i} V(h + az_0^j - (h + az_0^j)_{|\mathbf{p}_i}) \neq T_{\mathbf{p}_i} V(z_0 - z_0(\mathbf{p}_i))$ and $\mathbf{p}_i \notin \Sigma h$. Hence,

$$T_{\mathbf{p}_i} V(h - h(\mathbf{p}_i), z_0 - z_0(\mathbf{p}_i)) = T_{\mathbf{p}_i} V(h + az_0^j - (h + az_0^j)_{|\mathbf{p}_i}, z_0 - z_0(\mathbf{p}_i)) =$$

$$T_{\mathbf{p}_i} V(h + az_0^j - (h + az_0^j)_{|\mathbf{p}_i}) \cap T_{\mathbf{p}_i} V(z_0 - z_0(\mathbf{p}_i)) \to \mathcal{T} \cap T_{\mathbf{p}} V(z_0)$$

where, by taking a subsequence, we may assume that $T_{\mathbf{p}_i} V(h - h(\mathbf{p}_i))$ approaches some hyperplane, T. As $\mathfrak{G}$ is a good stratification for h, $T_{\mathbf{p}} G \subseteq T$. Moreover, as $V(z_0)$ transversely intersects G,

$$T_{\mathbf{p}_i} V(h - h(\mathbf{p}_i), z_0 - z_0(\mathbf{p}_i)) \to T \cap T_{\mathbf{p}} V(z_0)$$

and thus $T \cap T_{\mathbf{p}} V(z_0) = \mathcal{T} \cap T_{\mathbf{p}} V(z_0)$. Therefore,

$$T_{\mathbf{p}} G \cap T_{\mathbf{p}} V(z_0) \subseteq T \cap T_{\mathbf{p}} V(z_0) = \mathcal{T} \cap T_{\mathbf{p}} V(z_0) \subseteq \mathcal{T}. \ \square$$

Corollary 4.18. *Let $i \geqslant 0$ and suppose $(z_0, \ldots, z_i)$ is prepolar for h at the origin. If j is such that*

$$(*) \qquad \dim_{\mathbf{0}} \Gamma^{k+1}_{h,\mathbf{z}} \cap V\left(\frac{\partial h}{\partial z_0} + jaz_0^{j-1}\right) \cap V(z_1, \ldots, z_k) \leqslant 0$$

for all k with $0 \leqslant k \leqslant i$, then $(z_1, \ldots, z_i)$ is prepolar for $h + az_0^j$ at $\mathbf{0}$.

Proof. When $k = 0$, $(*)$ yields $\dim_{\mathbf{0}}\Gamma^{k+1}_{h,\mathbf{z}} \cap V\left(\frac{\partial h}{\partial z_0} + jaz_0^{j-1}\right) = 0$ and so, as sets,

$$\Sigma(h + az_0^j) = V\left(\frac{\partial h}{\partial z_0} + jaz_0^{j-1}\right) \cap V\left(\frac{\partial h}{\partial z_1}, \dots, \frac{\partial h}{\partial z_n}\right) =$$

$$V\left(\frac{\partial h}{\partial z_0} + jaz_0^{j-1}\right) \cap (\Sigma h \cup \Gamma^1_{h,\mathbf{z}}) =$$

$$(\Sigma h \cap V(z_0)) \cup \left(\Gamma^1_{h,\mathbf{z}} \cap V\left(\frac{\partial h}{\partial z_0} + jaz_0^{j-1}\right)\right) = \Sigma h \cap V(z_0).$$

Thus, the hypothesis of 4.17 is satisfied and we apply it; this leaves us with only the problem of showing that each successive hyperplane slice transversely intersects the smooth part, i.e. as germs of sets at the origin, for all k with $0 \leqslant k \leqslant i$,

$$\Sigma(h + az_0^j{}_{|V(z_1,\dots,z_k)}) = V\left(z_1, \dots, z_k, \frac{\partial h}{\partial z_0} + jaz_0^{j-1}\right) \cap V\left(\frac{\partial h}{\partial z_{k+1}}, \dots, \frac{\partial h}{\partial z_n}\right) =$$

$$V\left(z_1, \dots, z_k, \frac{\partial h}{\partial z_0} + jaz_0^{j-1}\right) \cap (\Sigma h \cup \Gamma^{k+1}_{h,\mathbf{z}}) =$$

$$(\Sigma h \cap V(z_0, \dots, z_k)) \cup \left(\Gamma^{k+1}_{h,\mathbf{z}} \cap V\left(\frac{\partial h}{\partial z_0} + jaz_0^{j-1}\right) \cap V(z_1, \dots, z_k)\right)$$

which, by $(*)$, equals $\Sigma h \cap V(z_0, \dots, z_k)$. □

Proposition 4.19. *Let $i \geqslant 0$ and suppose $(z_0, \dots, z_i)$ is prepolar for h at the origin. Then, for all large j,*

$$\dim_{\mathbf{0}}\Gamma^{k+1}_{h,\mathbf{z}} \cap V\left(\frac{\partial h}{\partial z_0} + jaz_0^{j-1}\right) \cap V(z_1, \dots, z_k) \leqslant 0$$

for all k with $0 \leqslant k \leqslant i$ and so $(z_1, \dots, z_i)$ is prepolar for $h + az_0^j$ at $\mathbf{0}$.

Proof. As $(z_0, \dots, z_i)$ is prepolar for h, we may apply 1.28 to conclude that $\gamma^{k+1}_{h,\mathbf{z}}(\mathbf{0})$ exists for all k with $0 \leqslant k \leqslant i$, i.e.

$$\dim_{\mathbf{0}}\Gamma^{k+1}_{h,\mathbf{z}} \cap V(z_0, z_1, \dots, z_k) \leqslant 0.$$

It follows immediately that

$$\dim_{\mathbf{0}}\Gamma^{k+1}_{h,\mathbf{z}} \cap V(z_1, \dots, z_k) \leqslant 1.$$

Therefore,

$$\dim_{\mathbf{0}}\Gamma^{k+1}_{h,\mathbf{z}} \cap V\left(\frac{\partial h}{\partial z_0} + jaz_0^{j-1}\right) \cap V(z_1, \dots, z_k) \nleqslant 0$$

if and only if $V\left(\frac{\partial h}{\partial z_0} + jaz_0^{j-1}\right)$ contains a component of $\Gamma^{k+1}_{h,\mathbf{z}} \cap V(z_1, \dots, z_k)$. But, if the same component of $\Gamma^{k+1}_{h,\mathbf{z}} \cap V(z_1, \dots, z_k)$ were contained in both $V\left(\frac{\partial h}{\partial z_0} + j_1az_0^{j_1-1}\right)$ and $V\left(\frac{\partial h}{\partial z_0} + j_2az_0^{j_2-1}\right)$ for $j_1 \neq j_2$, then z_0 would have to equal 0 along that component – a contradiction, as $\dim_{\mathbf{0}}W \cap V(z_0) \leqslant 0$. The conclusion follows. □

Chapter 5. LÊ NUMBERS AND HYPERPLANE ARRANGEMENTS

The Plücker formula of Corollary 4.7 states: let h be a homogeneous polynomial of degree d in $n+1$ variables, let $s = \dim_{\mathbf{0}}\Sigma h$, and suppose that $\lambda^i_{h,\mathbf{z}}(\mathbf{0})$ exists for all $i \leqslant s$. Then, $\sum_{i=0}^{s}(d-1)^i\lambda^i_{h,\mathbf{z}}(\mathbf{0}) = (d-1)^{n+1}$.

This formula allows us to calculate the Lê numbers for a central hyperplane arrangement in a purely combinatorial manner from the lattice of flats of the arrangement (see [**O-T**] and below). It was experimentally observed by D. Welsh and G. Ziegler that there was a fairly trivial relationship between the Lê numbers of the arrangement and the Möbius function (again, see [**O-T**] and below). This relationship generalizes to matroid-based polynomial identities (see [**MSSVWZ**]).

In this chapter, we give the combinatorial characterization of the Lê numbers for central hyperplane arrangements and prove the relation between the Lê numbers and the Möbius function.

A *central hyperplane arrangement in* $\mathbb{C}^{n+1}$ is simply the zero-locus of an analytic function $h : \mathbb{C}^{n+1} \to \mathbb{C}$ where h is a product of d linear forms on $\mathbb{C}^{n+1}$ (here, we are not necessarily assuming that the forms are distinct). Though this may appear to be fairly trivial as a hypersurface singularity, this apparent simplicity is deceiving – the study of hyperplane arrangements is quite complex and touches on many areas of mathematics (see, for instance, [**O-R**], [**O-S**],[**O-T**]).

Example 5.1. Suppose we have such an h. In this case, $V(h)$ equals the union of hyperplanes, $\{H_i\}_{i\in I}$, where I is the indexing set $\{1,\ldots,d'\}$, each H_i occurs with some multiplicity $m_i :=$ mult H_i, and $\sum m_i = d$ (in particular, if h is reduced, then each $m_i = 1$ and $d' = d$).

There is an obvious good, Whitney stratification of $V(h)$ obtained from the "flats" of the hyperplane arrangement; the collection of flats is given by $\{w_J\}_{J\subsetneq I}$, where

$$w_J := \bigcap_{i\in J} H_i.$$

If we now take the stratification $\{S_J\}_{J\subsetneq I}$, where

$$S_J = w_J - \bigcup_{J\subsetneqq K} w_K,$$

then clearly h is analytically trivial along the strata, and therefore one has trivially a Whitney stratification. In words, the strata are intersections of the hyperplanes minus smaller intersections of hyperplanes.

We wish to calculate the Lê numbers of h at the origin with respect to generic coordinates $\mathbf{z}$. As h is analytically trivial along the strata, it is easy to see that, as sets, the Lê cycles are given by the unions of the flats of correct dimension. Hence, as cycles, for all k,

$$\Lambda^k_{h,\mathbf{z}} = \sum_{\dim S_J = k} a_J [w_J]$$

for some a_J. By 1.21, a_J may be calculated by taking any $\mathbf{p} \in S_J$ and a normal slice N to S_J in $\mathbb{C}^{n+1}$ at $\mathbf{p}$, and then $a_J = \lambda^0_{h_{|N}}(\mathbf{p})$, where we use generic coordinates. After a translation to make the point $\mathbf{p}$ the origin, we see that $h_{|N}$ at $\mathbf{p}$ is again (up to multiplication by units) a product of linear forms of degree $e_J := \sum_{i \in J} m_i$.

Therefore, we may use 4.7 to calculate the Lê numbers of h at the origin by a downward induction on the dimension of the flats. (In the following, it looks nicer if we suppress the subscripts.) We denote a hyperplane in the arrangement by H, a flat by w or v, and define

$$e(w) := \sum_{w \subseteq H} \text{mult } H.$$

Next, we define the *vanishing Möbius function*, η, by downward induction on the dimension of the flats. For a hyperplane, H, in the arrangement, define

$$\eta(H) := \text{mult } H - 1;$$

for a smaller dimensional flat, w, 4.7 tells us that we need

$$\eta(w) := (e(w) - 1)^{n+1-\dim w} - \sum_{v \supsetneq w} \eta(v) \cdot (e(w) - 1)^{\text{codim}_v w}.$$

This equality is equivalent to

$$\sum_{v \supseteq w} (e(w) - 1)^{\dim v} \eta(v) = (e(w) - 1)^{n+1}.$$

Finally, having calculated the vanishing Möbius function, one has that, for all i,

$$\lambda^i_{h,\mathbf{z}}(\mathbf{0}) = \sum_{\dim w = i} \eta(w).$$

By 3.3, knowing the Lê numbers of the hyperplane arrangement gives us the Euler characteristic of the Milnor fibre together with Morse inequalities on the Betti numbers. (Another method for computing the Euler characteristic of the Milnor fibre from the data provided by the containment relations among the flats, i.e. by knowing the *intersection lattice*, is given in [**O-T**].)

Example 5.2. We wish to see what the above method gives us in the case of a generic central arrangement of d hyperplanes in $\mathbb{C}^{n+1}$ (see [**O-R**]). Here, "generic" means as generic as possible considering that all the hyperplanes pass through the origin – that is, each hyperplane occurs with multiplicity 1, and if w is a flat of dimension k, and $k \neq 0$, then w is the intersection of precisely $n+1-k$ hyperplanes of the arrangement; in terms of the above discussion, this says that if $w \neq \mathbf{0}$, then $e(w) = n+1-k$. We assume that $d > n+1$ for, otherwise, after a change of coordinates, $h = z_0 z_1 \dots z_{d-1}$ and the Milnor fibre is diffeomorphic to the $(d-1)$-fold product of $\mathbb{C}^*$'s.

For a generic arrangement, it is easy to see that for all j-dimensional flats $w \neq \mathbf{0}$, the number of k-dimensional flats containing w is given by $\binom{n+1-j}{k-j}$, provided that $k \geqslant j$. One also knows that, if $k \geqslant 1$, then the number of k-dimensional flats containing the origin is given by $\binom{d}{n+1-k}$. This is all the information that one needs to calculate the vanishing Möbius function, η, from the formula

$$\eta(w) := (e(w)-1)^{n+1-\dim w} - \sum_{v \supsetneq w} \eta(v) \cdot (e(w)-1)^{\mathrm{codim}_v w}$$

together with the fact that for all hyperplanes, H, in the arrangement we have $\eta(H) = 0$.

It is an amusing exercise to prove that this implies that, if $\dim w = j \neq 0$, then $\eta(w) = n-j$. Alternatively, this also follows from Example 2.8. (The above is the inductive proof of the formula of 2.8 that is referred to in that example.)

Therefore, for a generic central arrangement of d hyperplanes in $\mathbb{C}^{n+1}$, we have with respect to generic coordinates

$$\lambda^n_h(\mathbf{0}) = 0,$$

$$\lambda^{n-1}_h(\mathbf{0}) = \sum_{\dim w = n-1} \eta(w) = \binom{d}{2}(1),$$

$$\vdots$$

$$\lambda^i_{h,\mathbf{z}}(\mathbf{0}) = \sum_{\dim w = i} \eta(w) = \binom{d}{n+1-i}(n-i),$$

$$\vdots$$

$$\lambda^1_h(\mathbf{0}) = \sum_{\dim w = 1} \eta(w) = \binom{d}{n}(n-1).$$

So, finally,

$$\lambda^0_h(\mathbf{0}) = (d-1)^{n+1} - \sum_{i=1}^{n} (d-1)^i \lambda^i_{h,\mathbf{z}}(\mathbf{0}) =$$

$$(d-1)^{n+1} - \sum_{i=1}^{n}(d-1)^i\binom{d}{n+1-i}(n-i) =$$
$$(d-1)\binom{d-1}{n},$$

where the last equality is an exercise in combinatorics.

Now, by our earlier work, since we know the Lê numbers, we know the Euler characteristic of the Milnor fibre, $F_{h,0}$, together with Morse inequalities on the Betti numbers, $b_i(F_{h,0})$. But, in this special case, it is not difficult to obtain the Betti numbers precisely.

By an observation of D. Cohen [**Co**], if $d > n+1$, a generic central arrangement of d hyperplanes in $\mathbb{C}^{n+1}$ is obtained by taking repeated hyperplane sections of a generic hyperplane arrangement of d hyperplanes in $\mathbb{C}^d$. It follows that for $i \leqslant n-1$, $b_i(F_{h,0}) = \binom{d-1}{i}$. Therefore, we have only to calculate $b_n(F_{h,0})$; but, since we know the Euler characteristic, this is easy, and we find – after some more combinatorics- that

$$b_n(F_{h,0}) = (d-n)\binom{d-1}{n},$$

which agrees with the results of [**Co**] and [**O-R**].

Note that the Morse inequalities of 3.3 can be far from equalities; for instance, the two easiest inequalities are

$$(d-n)\binom{d-1}{n} = b_n(F_{h,0}) \leqslant \lambda^0_{h,\mathbf{z}}(\mathbf{0}) = (d-1)\binom{d-1}{n}$$

and

$$d-1 = b_1(F_{h,0}) \leqslant \lambda^{n-1}_{h,\mathbf{z}}(\mathbf{0}) = \binom{d}{2} = \frac{d(d-1)}{2}.$$

Now, we wish to describe the relation between the Lê numbers of a central arrangement and the Möbius function – this is the result which is generalized in [**MSSVWZ**].

Let h be the product of d distinct linear forms on $\mathbb{C}^{n+1}$, so that each hyperplane in the arrangement $V(h)$ occurs with multiplicity 1. Let $\mathcal{A}$ denote the collection of hyperplanes which are components of $V(h)$. We use the variable H to denote hyperplanes in $\mathcal{A}$. We use the letters v and w to denote flats of arbitrary dimension. Finally, in agreement with our notation in 5.1, let $e_{\mathcal{A}}(v) =$ the number of hyperplanes of $\mathcal{A}$ which contain the flat v.

As we saw in 5.1 and 5.2, the Lê numbers of a central hyperplane arrangement can be described in terms of a function $\eta_{\mathcal{A}}$ defined inductively on the flats by: for all $H \in \mathcal{A}$, $\eta_{\mathcal{A}}(H) = 0$, and for all flats w,

$$\sum_{w \subseteq v}(e_{\mathcal{A}}(w)-1)^{\dim v}\eta_{\mathcal{A}}(v) = (e_{\mathcal{A}}(w)-1)^{n+1}.$$

The *Möbius function*, $\mu_{\mathcal{A}}$, on $\mathcal{A}$ (see [O-T]) is defined inductively on the flats by: $\mu_{\mathcal{A}}(\mathbb{C}^{n+1}) = 1$ and for all flats $v \subsetneq w$,

$$\sum_{\substack{\text{flats } u \\ v \subseteq u \subseteq w}} \mu_{\mathcal{A}}(u) = 0.$$

Here, we subscript by η, e, and μ by $\mathcal{A}$ because our proof is by induction on the ambient dimension, and the inductive step requires slicing $\mathcal{A}$ by hyperplanes, N, **not** contained in $\mathcal{A}$. This will produce new arrangements inside the ambient space N. So it is important that we indicate which arrangement is under consideration.

More notation now, related to the slicing. We will be taking two kinds of hyperplane slices. N will denote a prepolar hyperplane slice through the origin in $\mathbb{C}^{n+1}$, i.e. a hyperplane slice which contains no flats of $\mathcal{A}$ other than the origin. We will also use normal slices to the one-flats; if v is a one-dimensional flat and $p_v \in v - 0$, N_v will denote a normal slice to v at p_v – that is, N_v is a hyperplane in $\mathbb{C}^{n+1}$ which transversely intersects v at p_v. We use $\mathcal{A} \cap N$ to denote the obvious induced arrangement in N (which is identified with $\mathbb{C}^n$). The arrangement $\mathcal{A} \cap N_v$ is considered as a central arrangement where p_v becomes the origin and all hyperplanes not containing p_v are ignored. Note that the number of hyperplanes in the arrangement $\mathcal{A} \cap N_v$ is $e_{\mathcal{A}}(v)$.

An arrangement is *essential* provided that the origin is a flat of the arrangement (hence, the arrangement is not trivially a product).

What we want to show is that, if $\mathcal{A}$ is a an essential, central hyperplane arrangement, then

$$\eta_{\mathcal{A}}(\mathbf{0}) = (d-1)(-1)^{n+1}\mu_{\mathcal{A}}(\mathbf{0}) = (d-1)|\mu_{\mathcal{A}}(\mathbf{0})|.$$

To induct, we will first need the following three easy lemmas on η, μ, which describe the effects of slicing. We leave the first two as exercises using the inductive definitions of $\eta_{\mathcal{A}}$ and$\mu_{\mathcal{A}}$ given above. However, we prove the third.

Lemma 5.3.

$$\eta_{\mathcal{A}\cap N}(\mathbf{0}) = \frac{\eta_{\mathcal{A}}(\mathbf{0})}{d-1} + \sum_{\dim v = 1} \eta_{\mathcal{A}}(v)$$

and, if v is a one-dimensional flat,

$$\eta_{\mathcal{A}\cap N_v}(p_v) = \eta_{\mathcal{A}}(v).$$

Lemma 5.4.

$$\mu_{\mathcal{A}\cap N}(\mathbf{0}) = -\sum_{\dim v \geq 2} \mu_{\mathcal{A}}(v).$$

and, if v is a one-dimensional flat,

$$\mu_{\mathcal{A}\cap N_v}(p_v) = \mu_{\mathcal{A}}(v).$$

Lemma 5.5.

$$d\mu_{\mathcal{A}}(\mathbf{0}) + \sum_{\dim v=1} (d - e_{\mathcal{A}}(v))\mu_{\mathcal{A}}(v) = 0.$$

Proof. By one of Weisner's formulas (see Lemma 2.40 of [**O-T**]), for all $H \in \mathcal{A}$,

$$\sum_{v\cap H=\mathbf{0}} \mu_{\mathcal{A}}(v) = 0.$$

Hence,

$$0 = \sum_{H}\left[\mu_{\mathcal{A}}(\mathbf{0}) + \sum_{\substack{\dim v=1\\ v\not\subseteq H}} \mu_{\mathcal{A}}(v)\right] = d\mu_{\mathcal{A}}(\mathbf{0}) + \sum_{\dim v=1} (d - e(v))\mu_{\mathcal{A}}(v). \square$$

Now, we can prove

Theorem 5.6. *If $\mathcal{A}$ is a an essential, central hyperplane arrangement consisting of d hyperplanes in $\mathbb{C}^{n+1}$, then*

$$\eta_{\mathcal{A}}(\mathbf{0}) = (d-1)(-1)^{n+1}\mu_{\mathcal{A}}(\mathbf{0}) = (d-1)|\mu_{\mathcal{A}}(\mathbf{0})|.$$

Proof. The proof is by induction on the ambient dimension. The formula is stupidly true when $n = 1$.

Now, suppose the formula is true for ambient dimension n. Then, $\eta_{\mathcal{A}\cap N}(\mathbf{0}) = (d-1)(-1)^n\mu_{\mathcal{A}\cap N}(\mathbf{0})$. Hence, by Lemmas 5.3 and 5.4,

$$\frac{\eta_{\mathcal{A}}(\mathbf{0})}{d-1} + \sum_{\dim v=1} \eta_{\mathcal{A}}(v) = (d-1)(-1)^n\left(-\sum_{\dim v\geqslant 2} \mu_{\mathcal{A}}(v)\right).$$

This gives us

$$\frac{\eta_{\mathcal{A}}(\mathbf{0})}{d-1} + \sum_{\dim v=1} \eta_{\mathcal{A}\cap N_v}(p_v) = (d-1)(-1)^{n+1}\sum_{\dim v\geqslant 2} \mu_{\mathcal{A}}(v).$$

Using our inductive hypothesis again, we have

$$\frac{\eta_{\mathcal{A}}(0)}{d-1} + \sum_{\dim v=1} (e_{\mathcal{A}}(v)-1)(-1)^n \mu_{\mathcal{A}\cap N_v}(p_v) = (d-1)(-1)^{n+1} \sum_{\dim v\geqslant 2} \mu_{\mathcal{A}}(v).$$

Therefore,

$$\frac{\eta_{\mathcal{A}}(0)}{d-1} + \sum_{\dim v=1} (e_{\mathcal{A}}(v)-1)(-1)^n \mu_{\mathcal{A}}(v) = (d-1)(-1)^{n+1} \sum_{\dim v\geq 2} \mu_{\mathcal{A}}(v)$$

and so

$$\eta_{\mathcal{A}}(0) = (d-1)(-1)^{n+1}\left[(d-1)\sum_{\dim v\geq 2} \mu_{\mathcal{A}}(v) + \sum_{\dim v=1} (e_{\mathcal{A}}(v)-1)\mu_{\mathcal{A}}(v)\right] =$$

$$(d-1)(-1)^{n+1}\left[(d-1)\Big(-\mu_{\mathcal{A}}(0) - \sum_{\dim v=1}\mu_{\mathcal{A}}(v)\Big) + \sum_{\dim v=1} (e_{\mathcal{A}}(v)-1)\mu_{\mathcal{A}}(v)\right]$$

$$= (d-1)(-1)^{n+1}\left[\mu_{\mathcal{A}}(0) - d\mu_{\mathcal{A}}(0) - \sum_{\dim v=1}(d-e_{\mathcal{A}}(v))\mu_{\mathcal{A}}(v)\right].$$

Now apply Lemma 5.5. □

Our inductive proof given above is somewhat unsatisfactory, for it gives us no geometric insight as to why the theorem is true. Should there be a geometric explanation for the identity in 5.6? Probably so. The result of Orlik and Solomon [**O-S**] is that $|\mu_{\mathcal{A}}(0)|$ is the $(n+1)$-st Betti number of the complement of the arrangement $\mathcal{A}$ in $\mathbb{C}^{n+1}$. How this could be used to prove 5.6 still escapes us.

Chapter 6. THOM'S a_f CONDITION

In this chapter, we will use the Lê numbers to provide conditions under which a submanifold of affine space satisfies Thom's a_f condition with respect to the ambient stratum. Let us recall the definition of the a_f condition (see [**Mat**]).

Definition 6.1. Let $\mathcal{V}$ be an open subset of some affine space, let $f : \mathcal{V} \to \mathbb{C}$ be an analytic function, and let M be a submanifold of $\mathcal{V}$. *Thom's a_f condition is satisfied* between $\mathcal{V} - \Sigma f$ and M (or *along* M) if, whenever $\mathbf{p}_i \in \mathcal{V} - \Sigma f$, $\mathbf{p}_i \to \mathbf{p} \in M$, and $T_{\mathbf{p}_i}V(f - f(\mathbf{p}_i)) \to \mathcal{T}$, then $T_{\mathbf{p}}M \subseteq \mathcal{T}$.

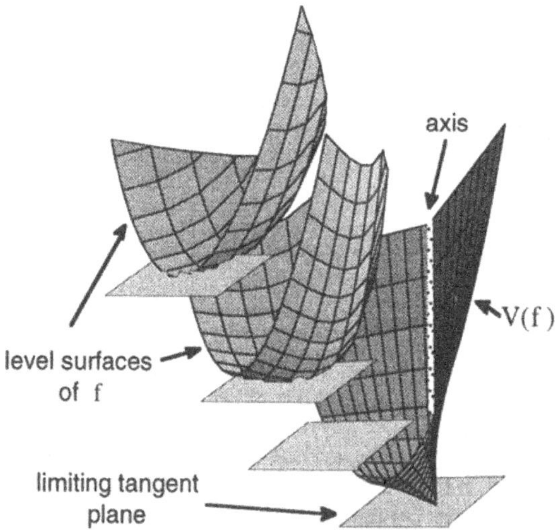

Figure 6.2. Failure of a_f along an axis

We are about to prove five different results – all of the form: Lê numbers constant $\Rightarrow$ Thom's a_f condition. The order in which we must prove these results is interesting.

First, we give a proof of Lê and Saito's result that a constant Milnor number at the origin in a one-parameter family implies the Thom condition along the parameter axis. We then use Lê and Saito's result, combined with the generalized Lê-Iomdine formulas, to prove that the constancy of the Lê numbers at the origin in a one-parameter family implies the Thom condition along the parameter axis. We use this parameterized version to prove a non-parameterized version: if the Lê numbers of a single function are constant along a submanifold, then Thom's condition is satisfied along the submanifold. This non-parameterized version allows us to prove a multi-parameter version of Lê and Saito's result: if we have a family of isolated hypersurface singularities with constant Milnor number parameterized along a submanifold, then that submanifold satisfies Thom's a_f condition with respect to the ambient stratum. Finally, we use this last result to prove our best result – the multi-parameter version of the Lê number result above: if we have a family of hypersurface singularities with constant Lê numbers parameterized along a submanifold, then that submanifold satisfies Thom's a_f condition with respect to the ambient stratum.

In all of the results described above, it is extremely important that **our assumptions on the genericity of the coordinate system will be solely that the Lê numbers exist**. This is a dimensional requirement which is very easy to check. This should be contrasted with the results of [**H-M**] and [**HMS**].

First, we need a well-known lemma.

Lemma 6.3. *Let $\overset{\circ}{\mathbb{D}}$ be an open disc about the origin in $\mathbb{C}$, let $\mathcal{U}$ be an open neighborhood of the origin in $\mathbb{C}^{n+1}$, and let $f : (\overset{\circ}{\mathbb{D}} \times \mathcal{U}, \overset{\circ}{\mathbb{D}} \times \mathbf{0}) \to (\mathbb{C}, 0)$ be an analytic function; we write f_t for the function defined by $f_t(\mathbf{z}) := f(t, \mathbf{z})$. Suppose that* $\dim_{\mathbf{0}}\Sigma f_0 = 0$.

Then, for all small t, the Milnor number of f_t at the origin is independent of t if and only if there exists an open neighborhood, $\mathcal{W}$, of the origin in $\overset{\circ}{\mathbb{D}} \times \mathcal{U}$ such that $\mathcal{W} \cap V\left(\frac{\partial f}{\partial z_0}, \dots, \frac{\partial f}{\partial z_n}\right) = \mathcal{W} \cap (\mathbb{C} \times \{\mathbf{0}\})$.

Proof. There are many proofs of this fact. We shall use intersection numbers.

The Milnor number of f_0 at the origin, $\mu_{\mathbf{0}}(f_0)$, equals the multiplicity of the origin in the cycle $\left[V\left(\frac{\partial f_0}{\partial z_0}, \dots, \frac{\partial f_0}{\partial z_n}\right)\right]$. Because f_0 has an isolated critical point at the origin, $V\left(t, \frac{\partial f}{\partial z_0}, \dots, \frac{\partial f}{\partial z_n}\right)$ is a local complete intersection, and so we have an equality of cycles

$$\left[V\left(\frac{\partial f_0}{\partial z_0}, \dots, \frac{\partial f_0}{\partial z_n}\right)\right] = [V(t)] \cdot \left[V\left(\frac{\partial f}{\partial z_0}, \dots, \frac{\partial f}{\partial z_n}\right)\right].$$

Therefore,

$$\mu_{\mathbf{0}}(f_0) =$$

$$\left(V(t) \cdot V\left(\frac{\partial f}{\partial z_0}, \dots, \frac{\partial f}{\partial z_n}\right)\right)_{\mathbf{0}} = \sum_{\mathbf{p} \in \overset{\circ}{B}_\epsilon} \left(V(t-\eta) \cdot V\left(\frac{\partial f}{\partial z_0}, \dots, \frac{\partial f}{\partial z_n}\right)\right)_{\mathbf{p}}$$

$$= \mu_{\mathbf{0}}(f_\eta) + R,$$

where $\overset{\circ}{B}_\epsilon$ is a sufficiently small open ball around the origin in $\mathbb{C}^{n+2}$, η is chosen small with respect to ϵ – that is, there exists $\delta_\epsilon > 0$ such that we may use any η satisfying $0 < |\eta| < \delta_\epsilon$ – and R denotes the sum of the remaining terms, i.e. the terms coming from points $\mathbf{p}$ which are not in $\mathbb{C} \times \{0\}$. Note that the sum is actually finite since we are really summing over $\mathbf{p} \in \mathring{B}_\epsilon \cap V(t-\eta) \cap V\left(\frac{\partial f}{\partial z_0}, \ldots, \frac{\partial f}{\partial z_n}\right)$. As all the intersection numbers are non-negative, R being zero is equivalent to there being no remaining terms, i.e. equivalent to $(\eta, \mathbf{0})$ being the only point in $\mathring{B}_\epsilon \cap V(t-\eta) \cap V\left(\frac{\partial f}{\partial z_0}, \ldots, \frac{\partial f}{\partial z_n}\right)$.

The desired conclusion follows immediately, where the set $\mathcal{W}$ in the statement can be taken to be $\mathring{B}_\epsilon \cap (\mathring{\mathbb{D}}_{\delta_\epsilon} \times \mathcal{U})$. □

Recall now the result of Lê and Saito [**L-S**] as stated in the introduction.

Theorem 6.4 (Lê-Saito [**L-S**]). *Let $\mathring{\mathbb{D}}$ be an open disc about the origin in $\mathbb{C}$, let $\mathcal{U}$ be an open neighborhood of the origin in $\mathbb{C}^{n+1}$, and let $f : (\mathring{\mathbb{D}} \times \mathcal{U}, \mathring{\mathbb{D}} \times \mathbf{0}) \to (\mathbb{C}, 0)$ be an analytic function; we write f_t for the function defined by $f_t(\mathbf{z}) := f(t, \mathbf{z})$.*

Suppose that $\dim_{\mathbf{0}} \Sigma f_0 = 0$ *and that, for all small t, the Milnor number of f_t at the origin is independent of t. Then, $\mathring{\mathbb{D}} \times \{0\}$ satisfies Thom's a_f condition at the origin with respect to the ambient stratum, i.e. if $\mathbf{p}_i$ is a sequence of points in $\mathring{\mathbb{D}} \times \mathcal{U} - \Sigma f$ such that $\mathbf{p}_i \to \mathbf{0}$ and such that $T_{\mathbf{p}_i} V(f - f(\mathbf{p}_i))$ converges to some $\mathcal{T}$, then $\mathbb{C} \times \mathbf{0} = T_{\mathbf{0}}(\mathring{\mathbb{D}} \times \{0\}) \subseteq \mathcal{T}$.*

Proof. We begin by noting that the existence of good stratifications as given in Proposition 1.25 implies that Thom's a_f is satisfied, near the origin, by $\mathring{\mathbb{D}} \times \{0\} - \mathbf{0}$ with respect to the ambient stratum.

Now, consider the blow-up of $\mathring{\mathbb{D}} \times \mathcal{U}$ by the Jacobian ideal of f:

$$\begin{array}{ccc} & Bl_{J(f)}(\mathring{\mathbb{D}} \times \mathcal{U}) \subseteq (\mathring{\mathbb{D}} \times \mathcal{U}) \times \mathbb{P}^{n+1} & \\ & \pi_1 \swarrow \qquad \searrow \pi_2 & \\ \mathring{\mathbb{D}} \times \mathcal{U} & & \mathbb{P}^{n+1} \end{array}$$

We first wish to show that the fibre $\pi_1^{-1}(\mathbf{0})$ has dimension at most n.

The point $\mathbf{q} := [1 : 0 : \cdots : 0] \in \mathbb{P}^{n+1}$ corresponds to the hyperplane $V(t)$. As $\mu_{\mathbf{0}}(f_t)$ is independent of t, the lemma implies that, in a neighborhood of the origin, $\pi_1(\pi_2^{-1}(\mathbf{q})) \subseteq \mathring{\mathbb{D}} \times \{0\}$. However, as we noted above, the a_f condition holds generically on $\mathring{\mathbb{D}} \times \{0\}$. Therefore, near $\mathbf{0}$, either $\pi_1(\pi_2^{-1}(\mathbf{q}))$ is empty or consists only of the origin. But, the dimension of every component of $\pi_2^{-1}(\mathbf{q})$ is at least $\dim Bl_{J(f)}(\mathring{\mathbb{D}} \times \mathcal{U}) - \dim \mathbb{P}^{n+1} = n+2-(n+1) = 1$. Thus, $\mathbf{0} \notin \pi_1(\pi_2^{-1}(\mathbf{q}))$, i.e. $\mathbf{q} \notin \pi_2(\pi_1^{-1}(\mathbf{0}))$. It follows that $\pi_1^{-1}(\mathbf{0})$ is a proper subset of $\mathbb{P}^{n+1}$ and, hence, has dimension at most n.

But, every component of the exceptional divisor $E := \pi_1^{-1}(\Sigma f)$ has dimension $n+1$. Therefore, above an open neighborhood of the origin, E equals the topological closure of $E - \pi_1^{-1}(\mathbf{0})$, which is contained in $(\overset{\circ}{\mathbb{D}} \times \{0\}) \times (\{0\} \times \mathbb{P}^n)$ since the a_f condition holds generically on the t-axis. It follows that $\pi_2(\pi_1^{-1}(\mathbf{0})) \subseteq \{0\} \times \mathbb{P}^n$, i.e. that the a_f condition holds along $\overset{\circ}{\mathbb{D}} \times \{0\}$ at the origin. □

Our first generalization of the result of Lê and Saito is:

Theorem 6.5. *Let $\overset{\circ}{\mathbb{D}}$ be an open disc about the origin in $\mathbb{C}$, let $\mathcal{U}$ be an open neighborhood of the origin in $\mathbb{C}^{n+1}$, and let $f : (\overset{\circ}{\mathbb{D}} \times \mathcal{U}, \overset{\circ}{\mathbb{D}} \times \mathbf{0}) \to (\mathbb{C}, 0)$ be an analytic function; we write f_t for the function defined by $f_t(\mathbf{z}) := f(t, \mathbf{z})$.*

Let $s = \dim_{\mathbf{0}} \Sigma f_0$. Suppose that, for all small t, for all i with $0 \leqslant i \leqslant s$, $\lambda^i_{f_t,\mathbf{z}}(\mathbf{0})$ is defined and is independent of t. Then, $\overset{\circ}{\mathbb{D}} \times \mathbf{0}$ satisfies Thom's a_f condition at the origin with respect to the ambient stratum, i.e. if $\mathbf{p}_i$ is a sequence of points in $\overset{\circ}{\mathbb{D}} \times \mathcal{U} - \Sigma f$ such that $\mathbf{p}_i \to \mathbf{0}$ and such that $T_{\mathbf{p}_i} V(f - f(\mathbf{p}_i))$ converges to some $\mathcal{T}$, then $\mathbb{C} \times \mathbf{0} = T_{\mathbf{0}}(\overset{\circ}{\mathbb{D}} \times \mathbf{0}) \subseteq \mathcal{T}$.

Proof. The proof is by induction on s. For $s = 0$, the theorem is exactly that of Lê and Saito in 6.4.

Now, suppose that $s \geqslant 1$ and that, for all small t, for all i with $0 \leqslant i \leqslant s$, $\lambda^i_{f_t,\mathbf{z}}(\mathbf{0})$ is defined and is independent of t, but that there exists a sequence $\mathbf{p}_i$ of points in $\overset{\circ}{\mathbb{D}} \times \mathcal{U} - \Sigma f$ such that $\mathbf{p}_i \to \mathbf{0}$, such that $T_{\mathbf{p}_i} V(f - f(\mathbf{p}_i))$ converges to some $\mathcal{T}$, and $T_{\mathbf{0}}(\overset{\circ}{\mathbb{D}} \times \mathbf{0}) \not\subseteq \mathcal{T}$.

As the collection of such limiting $\mathcal{T}$ is analytic, we may apply the curve selection lemma (see [**Lo**]) to conclude that there exists a real analytic curve

$$\alpha : [0, \epsilon) \to \{\mathbf{0}\} \cup (\overset{\circ}{\mathbb{D}} \times \mathcal{U} - \Sigma f)$$

such that $\alpha(u) = \mathbf{0}$ if and only if $u = 0$ and such that

$$\lim_{u \to 0} T_{\alpha(u)} V(f - f(\alpha(u))) = \mathcal{T}.$$

As α is real analytic, it is trivial to show that, for all large j,

$$\lim_{u \to 0} \frac{\operatorname{grad}(f + z_0^j)_{|\alpha(u)}}{|\operatorname{grad}(f + z_0^j)_{|\alpha(u)}|} = \lim_{u \to 0} \frac{\operatorname{grad}(f)_{|\alpha(u)}}{|\operatorname{grad}(f)_{|\alpha(u)}|}.$$

Therefore, for all large j, the family $f_t + z_0^j$ also has $\mathcal{T}$ as a limit to level hypersurfaces, i.e. $\overset{\circ}{\mathbb{D}} \times \mathbf{0}$ does not satisfy the $a_{f+z_0^j}$ condition at the origin with respect to the ambient stratum.

However, $\lambda^0_{f_t,\mathbf{z}}(\mathbf{0})$ is independent of t and, applying Theorem 4.5, if $j \geqslant 2 + \lambda^0_{f_t,\mathbf{z}}(\mathbf{0})$, then the family $f_t + z_0^j$ has Lê numbers independent of t, and $f_0 + z_0^j$ has a critical locus of dimension $s - 1$. Thus, our inductive hypothesis contradicts the previous paragraph. □

Corollary 6.6. *Let $h : \mathcal{U} \to \mathbb{C}$ be an analytic function on an open subset of $\mathbb{C}^{n+1}$, let $\mathbf{z} = (z_0, \ldots, z_n)$ be a linear choice of coordinates for $\mathbb{C}^{n+1}$, let M be an analytic submanifold of $V(h)$, let $\mathbf{q} \in M$, and let s denote $\dim_{\mathbf{q}}\Sigma h$.*

If, for each i such that $0 \leqslant i \leqslant s$, $\lambda^i_{h,\mathbf{z}}(\mathbf{p})$ is defined and is independent of $\mathbf{p}$, for all $\mathbf{p} \in M$ near $\mathbf{q}$, then M satisfies Thom's a_h condition at $\mathbf{q}$ with respect to the ambient stratum; that is, if $\mathbf{q}_i$ is a sequence of points in $\mathcal{U} - \Sigma h$ such that $\mathbf{q}_i \to \mathbf{q}$ and such that $T_{\mathbf{q}_i} V(h - h(\mathbf{q}_i))$ converges to some $\mathcal{T}$, then $T_{\mathbf{q}} M \subseteq \mathcal{T}$.

Proof. This follows from 6.5 by a fairly standard trick. Let $\mathbf{c}(t)$ be a smooth analytic path in M such that $\mathbf{c}(0) = \mathbf{q}$. If we can show that any limiting tangent plane, $\mathcal{T}$, contains the tangent to the image of $\mathbf{c}$ at $\mathbf{q}$, then we will be finished.

So, take such a $\mathbf{c}$, and suppose that we have a sequence of points, $\mathbf{q}_i$, in $\mathcal{U} - \Sigma h$ such that $\mathbf{q}_i \to \mathbf{q}$ and such that $T_{\mathbf{q}_i} V(h - h(\mathbf{q}_i)) \to \mathcal{T}$.

Define $f(t, \mathbf{z}) := h(\mathbf{z} + \mathbf{c}(t))$, and consider the sequence of points $(0, \mathbf{q}_i - \mathbf{q})$. If one now applies the theorem, the result is that $\mathbf{c}'(0) \subseteq \mathcal{T}$; we leave the details to the reader. □

Remark 6.7. It is important to note that, in 6.6, we only require that the coordinates are generic enough so that the Lê numbers are defined; we are not requiring that the coordinates are prepolar.

On the other hand, Corollary 6.6 tells us how we can obtain good stratifications: if we have an analytic stratification of $V(h)$ such that the Lê numbers are defined and constant along the strata, then the stratification is actually a good stratification. However, there is no guarantee that the coordinates used to define the Lê numbers are prepolar with respect to this good stratification.

Now we can prove the multi-parameter version of the result of Lê and Saito.

Theorem 6.8. *Let M be an open neighborhood of the origin in $\mathbb{C}^k$, let $\mathcal{U}$ be an open neighborhood of the origin in $\mathbb{C}^{n+1}$, and let $f : (M \times \mathcal{U}, M \times \mathbf{0}) \to (\mathbb{C}, 0)$ be an analytic function; we write $f_{\mathbf{t}}$ for the function defined by $f_{\mathbf{t}}(\mathbf{z}) := f(\mathbf{t}, \mathbf{z})$, where $\mathbf{t} \in M$ and $\mathbf{z} \in \mathcal{U}$.*

Suppose that $\dim_{\mathbf{0}}\Sigma f_{\mathbf{0}} = 0$ and that, for all $\mathbf{t}$ near the origin, the Milnor number of $f_{\mathbf{t}}$ at the origin is independent of $\mathbf{t}$. Then, $M \times \mathbf{0}$ satisfies Thom's a_f condition at the origin with respect to the ambient stratum, i.e. if $\mathbf{p}_i$ is a sequence of points in $M \times \mathcal{U} - \Sigma f$ such that $\mathbf{p}_i \to \mathbf{0}$ and such that $T_{\mathbf{p}_i} V(f - f(\mathbf{p}_i))$ converges to some $\mathcal{T}$, then $T_{\mathbf{0}}(M \times \mathbf{0}) \subseteq \mathcal{T}$.

Proof. If the constant value of the Milnor number is 0, then there is nothing to prove. Note, though, that if the constant value of the Milnor number is non-zero, then it follows from Sard's theorem that $M \times \mathbf{0} \subseteq \Sigma f$, i.e. the critical points of the $f_{\mathbf{t}}$ are not merely a result of critical points of the map $\mathbf{t}$ restricted to the smooth part of $V(f)$.

Let $\mathbf{a}$ be an element of M near the origin. The Milnor number of $f_{\mathbf{a}}$ at the origin satisfies the equality

$$\mu_0(f_{\mathbf{a}}) = \left[V\left(t_0 - a_0, \ldots, t_{k-1} - a_{k-1}, \frac{\partial f}{\partial z_0}, \ldots, \frac{\partial f}{\partial z_n}\right)\right]_{(\mathbf{a},\mathbf{0})}.$$

In particular,

$$\dim_{(\mathbf{a},\mathbf{0})} V\left(t_0 - a_0, \ldots, t_{k-1} - a_{k-1}, \frac{\partial f}{\partial z_0}, \ldots, \frac{\partial f}{\partial z_n}\right) = 0.$$

This immediately implies that $\dim_{(\mathbf{a},\mathbf{0})}\Sigma f \leqslant k$. Hence,

$$\left[V\left(\frac{\partial f}{\partial z_0}, \ldots, \frac{\partial f}{\partial z_n}\right)\right] = \Gamma^k_{f,(\mathbf{t},\mathbf{z})} + \Lambda^k_{f,(\mathbf{t},\mathbf{z})},$$

both $\gamma^k_{f,(\mathbf{t},\mathbf{z})}(\mathbf{a},\mathbf{0})$ and $\lambda^k_{f,(\mathbf{t},\mathbf{z})}(\mathbf{a},\mathbf{0})$ exist, and

$$\mu_0(f_{\mathbf{a}}) = \gamma^k_{f,(\mathbf{t},\mathbf{z})}(\mathbf{a},\mathbf{0}) + \lambda^k_{f,(\mathbf{t},\mathbf{z})}(\mathbf{a},\mathbf{0}).$$

As $\mu_0(f_{\mathbf{a}})$ is independent of $\mathbf{a}$, and both $\gamma^k_{f,(\mathbf{t},\mathbf{z})}(\mathbf{a},\mathbf{0})$ and $\lambda^k_{f,(\mathbf{t},\mathbf{z})}(\mathbf{a},\mathbf{0})$ are upper-semicontinuous as functions of $\mathbf{a}$, we conclude that both $\gamma^k_{f,(\mathbf{t},\mathbf{z})}(\mathbf{a},\mathbf{0})$ and $\lambda^k_{f,(\mathbf{t},\mathbf{z})}(\mathbf{a},\mathbf{0})$ are independent of $\mathbf{a}$.

This implies that $\gamma^k_{f,(\mathbf{t},\mathbf{z})}(\mathbf{a},\mathbf{0})$ is independent of $\mathbf{a}$ for $(\mathbf{a},\mathbf{0})$ in a k-dimensional component of Σf. But, $\Gamma^k_{f,(\mathbf{t},\mathbf{z})}$ cannot contain a component of Σf. Therefore, the constant value of $\gamma^k_{f,(\mathbf{t},\mathbf{z})}(\mathbf{a},\mathbf{0})$ for $\mathbf{a} \in M$ must be 0; that is, $\Gamma^k_{f,(\mathbf{t},\mathbf{z})}$ does not intersect $M \times \mathbf{0}$ near the origin.

But, all the lower-dimensional relative polar cycles are contained in $\Gamma^k_{f,(\mathbf{t},\mathbf{z})}$; thus, none of them hit $M \times \mathbf{0}$. This implies that $\lambda^i_{f,(\mathbf{t},\mathbf{z})}(\mathbf{a},\mathbf{0}) = 0$ for all $\mathbf{a} \in M$ and all i with $0 \leqslant i \leqslant k-1$. As we already saw that $\lambda^k_{f,(\mathbf{t},\mathbf{z})}(\mathbf{a},\mathbf{0})$ is independent of $\mathbf{a} \in M$, we see that all the Lê numbers of f are constant along $M \times \mathbf{0}$.

Now, apply Corollary 6.6. □

Finally, we have the multi-parameter generalization of the result of Lê and Saito, where the critical loci may have arbitrary dimension.

Theorem 6.9. *Let M be an open neighborhood of the origin in $\mathbb{C}^k$, let $\mathcal{U}$ be an open neighborhood of the origin in $\mathbb{C}^{n+1}$, and let $f : (M \times \mathcal{U}, M \times \mathbf{0}) \to (\mathbb{C}, 0)$*

be an analytic function; we write $f_{\mathbf{t}}$ *for the function defined by* $f_{\mathbf{t}}(\mathbf{z}) := f(\mathbf{t}, \mathbf{z})$, *where* $\mathbf{t} \in M$ *and* $\mathbf{z} \in \mathcal{U}$.

Let $s = \dim_{\mathbf{0}} \Sigma f_{\mathbf{0}}$. *Suppose that, for all small* $\mathbf{t}$, *for all* i *with* $0 \leqslant i \leqslant s$, $\lambda^i_{f_{\mathbf{t}}, \mathbf{z}}(\mathbf{0})$ *is defined and is independent of* $\mathbf{t}$. *Then,* $M \times \mathbf{0}$ *satisfies Thom's* a_f *condition at the origin with respect to the ambient stratum, i.e. if* $\mathbf{p}_i$ *is a sequence of points in* $M \times \mathcal{U} - \Sigma f$ *such that* $\mathbf{p}_i \to \mathbf{0}$ *and such that* $T_{\mathbf{p}_i} V(f - f(\mathbf{p}_i))$ *converges to some* $\mathcal{T}$, *then* $T_{\mathbf{0}}(M \times \mathbf{0}) \subseteq \mathcal{T}$.

Proof. The proof is by induction on s. To obtain 6.9 from 6.8, one follows word for word our derivation of 6.5 from 6.4. $\square$

Chapter 7. ALIGNED SINGULARITIES

In this chapter, we once again consider analytic functions $h : \mathcal{U} \to \mathbb{C}$. We wish to investigate those h for which the critical locus, Σh, is of a particularly nice form – a form which generalizes isolated singularities, smooth one-dimensional singularities (line singularities), and the singularities found in hyperplane arrangements.

The obvious generalization of merely requiring Σh to be smooth appears to be too general to yield nice results; what one would like is to put some restrictions on the subset of Σh where h fails to be "equisingular". For instance, in the case where Σh is smooth and 2-dimensional, any reasonable notion of equisingularity could fail on a subset of dimension at most one; a reasonable condition to impose is that this one-dimensional subset itself be smooth. Essentially this is what we require of an aligned singularity.

For convenience, throughout this section, we concentrate our attention on hypersurface germs at the origin.

Definition 7.1. If $h : (\mathcal{U}, \mathbf{0}) \to (\mathbb{C}, 0)$ is an analytic function, then an *aligned good stratification* for h at the origin is a good stratification for h at the origin in which the closure of each stratum of the singular set is smooth at the origin.

If such an aligned good stratification exists, we say that h has an *aligned singularity* at the origin.

If $\{S_\alpha\}$ is an aligned good stratification for h at the origin, then we say that a linear choice of coordinates, $\mathbf{z}$, is *an aligning set of coordinates* for $\{S_\alpha\}$ provided that for each i, $V(z_0, \dots, z_{i-1})$ transversely intersects the closure of each stratum of dimension $\geqslant i$ of $\{S_\alpha\}$ at the origin. Naturally, we say simply that a set of coordinates, $\mathbf{z}$, is *aligning for h* at the origin provided that there exists an aligned good stratification for h at the origin with respect to which $\mathbf{z}$ is aligning.

Note that, given an aligned singularity, aligning sets of coordinates are generic and prepolar.

Closely related to this notion is

Definition 7.2. If $h : (\mathcal{U}, \mathbf{0}) \to (\mathbb{C}, 0)$ is an analytic function on an open subset of $\mathbb{C}^{n+1}$, then a linear choice of coordinates, $\mathbf{z}$, for $\mathbb{C}^{n+1}$ is *pre-aligning for h* at the origin provided that for each Lê cycle, $\Lambda^i_{h,\mathbf{z}}$, and for each irreducible component, C, of $\Lambda^i_{h,\mathbf{z}}$ passing through the origin, the following conditions are satisfied:

i) $\dim_{\mathbf{0}} C = i$;

ii) C is smooth at the origin;

iii) $V(z_0, z_1, \ldots, z_{i-1})$ transversely intersects C at the origin.

Proposition 7.3. *If h has an aligned singularity at the origin, then for a generic linear choice of coordinates $\mathbf{z}$, $\mathbf{z}$ is prepolar for h at each point, $\mathbf{p}$, near the origin and, hence, for each such $\mathbf{p}$, the reduced Euler characteristic of the Milnor fibre of h at $\mathbf{p}$ is given by*

$$\tilde{\chi}(F_{h,\mathbf{p}}) = \sum_{i=0}^{s} (-1)^{n-i} \lambda^i_{h,\mathbf{p}}(\mathbf{0}).$$

Proof. One may simply choose $\mathbf{z}$ to be aligning. The Euler characteristic statement then follows at once from Theorem 3.3. □

Proposition 7.4. *Suppose that $\{S_\alpha\}$ is an aligned good stratification for h at the origin and that $\mathbf{z}$ is an aligning set of coordinates for $\{S_\alpha\}$ at the origin. Then, as germs of sets at the origin, for all i,*

$$\Lambda^i_{h,\mathbf{z}} \subseteq \bigcup_{\dim_{\mathbf{0}} S_\alpha = i} \overline{S}_\alpha.$$

Hence, $\mathbf{z}$ is a pre-aligning set of coordinates for h at the origin.

Proof. We proceed by induction on i. Note that, as the coordinates are prepolar at the origin, each of the Lê cycles has the proper dimension near the origin.

For $i = 0$, we must show that if the origin is in $\Lambda^0_{h,\mathbf{z}}$, then the origin is also a stratum. Suppose not.

Then, $\mathbf{0}$ is in some stratum S of dimension $\geqslant 1$ and, as $\mathbf{0} \in \Lambda^0_{h,\mathbf{z}}$, $\mathbf{0} \in \Gamma^1_{h,\mathbf{z}}$. As our coordinates are aligning, $V(z_0)$ transversely intersects $\overline{S}$ at the origin. This, however, is a contradiction, since $\mathbf{0} \in \Gamma^1_{h,\mathbf{z}}$ implies that there is a sequence of limiting tangent planes to level hypersurfaces which converges to $V(z_0)$ at the origin and, hence, $T_{\mathbf{0}}S$ should be contained in $V(z_0)$ since S is a good stratum.

Now, suppose that $i \geqslant 1$. As each $\Lambda^i_{h,\mathbf{z}}$ has the proper dimension, it clearly suffices to show that

$$\Lambda^i_{h,\mathbf{z}} - \{\mathbf{0}\} \subseteq \bigcup_{\dim_{\mathbf{0}} S_\alpha \leqslant i} \overline{S}_\alpha.$$

Suppose not; let $\mathbf{p}$ be such that

$$\mathbf{p} \in \Lambda^i_{h,\mathbf{z}} - \{\mathbf{0}\} \text{ and } \mathbf{p} \notin \bigcup_{\dim_{\mathbf{0}} S_\alpha \leqslant i} \overline{S}_\alpha.$$

Then, $V(z_0 - p_0)$ transversely intersects all strata of $\{S_\alpha\}$ near $\mathbf{p}$ and the coordinates $\tilde{\mathbf{z}} := (z_1, \ldots, z_n)$ are such that $V(z_1, \ldots, z_i)$ transversely intersects the closure of any i-dimensional stratum of the aligned good stratification $\{S_\alpha \cap V(z_0 - p_0)\}$ at all points near $\mathbf{p}$. Moreover, by Proposition 1.21, as sets,

$$\Lambda^{i-1}_{h_{|_{V(z_0-p_0)}},\tilde{\mathbf{z}}} = \Lambda^i_{h,\mathbf{z}} \cap V(z_0 - p_0).$$

Hence, at $\mathbf{p}$, $\Lambda^{i-1}_{h_{|_{V(z_0-p_0)}},\tilde{\mathbf{z}}}$ would not be contained in

$$\bigcup_{\dim_{\mathbf{0}} S_\alpha \cap V(z_0-p_0) \leqslant i-1} \overline{S_\alpha \cap V(z_0 - p_0)}.$$

But, this would contradict our inductive hypothesis. □

Remark 7.5. It is tempting to think that if $\mathbf{z}$ is a set of pre-aligning coordinates for h at the origin, then we can produce an aligned good stratification by considering the components of

$$\Lambda^i_{h,\mathbf{z}} - \bigcup_{j \leqslant i-1} \Lambda^j_{h,\mathbf{z}}.$$

This might seem reasonable in light of Corollary 6.6 and Remark 6.7. However, we see no reason for the higher Lê numbers to be constant along these proposed "strata".

We could define a more restricted class of singularities – *super aligned singularities* – by requiring the existence of an aligned good stratification in which, for each i with $0 \leqslant i \leqslant s := \dim_{\mathbf{0}} \Sigma h$, there is at most one connected stratum, say S^i, of dimension i and

$$S^0 \subseteq \overline{S^1} \subseteq \cdots \subseteq \overline{S^{s-1}} \subseteq \overline{S^s}.$$

It is easy to see that this is equivalent to the existence of a set of pre-aligning coordinates, $\mathbf{z}$, for h at the origin such that each $\Lambda^i_{h,\mathbf{z}}$ has a single smooth component at the origin and, as germs of sets at the origin,

$$\Lambda^0_{h,\mathbf{z}} \subseteq \Lambda^1_{h,\mathbf{z}} \subseteq \cdots \subseteq \Lambda^s_{h,\mathbf{z}}.$$

For such super aligned singularities, we obtain a good stratification for h by taking the stratification

$$\{\Lambda^{j+1}_{h,\mathbf{z}} - \Lambda^j_{h,\mathbf{z}}\}.$$

Our main interest in aligned singularities is due to

Proposition 7.6. *Suppose that h has an aligned s-dimensional singularity at the origin and that the coordinates $\mathbf{z}$ are aligning. Then, the Lê cycles and Lê numbers can be characterized topologically in the following inductive manner:*

as a set, $\Lambda^s_{h,\mathbf{z}}$ equals the union of the s-dimensional components of the singular set of h. To determine the Lê cycle, to each s-dimensional component, C of Σh, we assign the multiplicity $m_C = (-1)^{n-s}\tilde{\chi}(F_{h,\mathbf{p}})$ for generic $\mathbf{p} \in C$, where $F_{h,\mathbf{p}}$ denotes the Milnor fibre of h at $\mathbf{p}$ and $\tilde{\chi}$ is the reduced Euler characteristic. Moreover, for all $\mathbf{p} \in |\Lambda^s_{h,\mathbf{z}}|$, $\lambda^s_{h,\mathbf{z}}(\mathbf{p}) = \sum_{\mathbf{p}\in C} m_C$.

Now, suppose that we have defined the Lê numbers, $\lambda^i_{h,\mathbf{z}}(\mathbf{p})$ for all $i \geqslant k+1$ and for all $\mathbf{p}$ near the origin.

Then, as a set, $\Lambda^k_{h,\mathbf{z}}$ equals the closure of the k-dimensional components of the set of points $\mathbf{p} \in V(h)$ where

$$\tilde{\chi}(F_{h,\mathbf{p}}) \neq \sum_{i=k+1}^{s} (-1)^{n-i}\lambda^i_{h,\mathbf{z}}(\mathbf{p}).$$

The Lê cycle is defined by assigning to each irreducible component C of this set the multiplicity

$$m_C = (-1)^{n-k}\left(\tilde{\chi}(F_{h,\mathbf{p}}) - \sum_{i=k+1}^{s} (-1)^{n-i}\lambda^i_{h,\mathbf{z}}(\mathbf{p})\right),$$

for generic $\mathbf{p} \in C$. Finally, for all $\mathbf{p} \in |\Lambda^k_{h,\mathbf{z}}|$, we have $\lambda^k_{h,\mathbf{z}}(\mathbf{p}) = \sum_{\mathbf{p}\in C} m_C$.

Proof. As the aligning coordinates are prepolar at each point near the origin, this essentially follows from Theorem 3.3. However, in writing that

$$\lambda^k_{h,\mathbf{z}}(\mathbf{p}) = \sum_{\mathbf{p}\in C} m_C,$$

we are crucially using that the components of the Lê cycle are smooth and tranversely intersected by $V(z_0 - p_0, \ldots, z_{k-1} - p_{k-1})$. If this were not the case, the intersection multiplicities of $V(z_0 - p_0, \ldots, z_{k-1} - p_{k-1})$ with the components of the Lê cycles would enter the picture, and the characterization would no longer be purely topological. □

The following two corollaries are immediate:

Corollary 7.7. *If h has an aligned singularity at the origin, then all aligning coordinates $\mathbf{z}$ determine the same Lê cycles and Lê numbers.*

Corollary 7.8. *Let f and g be reduced, analytic germs with aligned singularities at the origin in $\mathbb{C}^{n+1}$. Let $\mathbf{z}$ and $\tilde{\mathbf{z}}$ be aligning sets of coordinates for f and g,* *respectively. If H is a local, ambient homeomorphism from the germ of V(f) at the origin to the germ of V(g) at the origin, then as germs of sets at the origin,*

$$H(\Lambda^i_{f,\mathbf{z}}) = \Lambda^i_{g,\tilde{\mathbf{z}}},$$

for all i, and for all $\mathbf{p}$ *near the origin in* $\mathbb{C}^{n+1}$,

$$\lambda^i_{f,\mathbf{z}}(\mathbf{p}) = \lambda^i_{g,\tilde{\mathbf{z}}}(H(\mathbf{p})),$$

for all i.

Now, we will give an amusing application of the results of this section. In [**Z**], Zariski conjectures that the multiplicity of a hypersurface at a point is an invariant of the local, ambient topological type of the hypersurface. A number of people have concentrated on the case of a one-parameter family of isolated singularities, but even this case has not been settled (however, for families of quasihomogeneous isolated singularities, the proof of the conjecture has been given by Greuel [**Gr**] and O'Shea [**O'S**]).

In our paper [**Mas7**], we prove a result which perhaps supplies a better place to look for counterexamples to the Zariski Multiplicity Conjecture; we prove, for families of hypersurfaces of dimension $\neq 2$, that the Zariski Multiplicity Conjecture is true for families of hypersurfaces with isolated singularities if and only if it is true for families of hypersurfaces with smooth one-dimensional critical loci. The results of this section allow us to generalize this.

Theorem 7.9. *The following are equivalent:*

i) for all $n \geqslant 3$, *the Zariski Multiplicity Conjecture is true for families of reduced analytic hypersurfaces* $f_t : (\mathcal{U}, \mathbf{0}) \to (\mathbb{C}, 0)$, *where* $\mathcal{U}$ *is an open subset of* $\mathbb{C}^{n+1}$ *and* $\dim_{\mathbf{0}}\Sigma f_t = 0$;

ii) for all $n \geqslant 3$, *there exists a k such that the Zariski Multiplicity Conjecture is true for families of reduced analytic hypersurfaces* $f_t : (\mathcal{U}, \mathbf{0}) \to (\mathbb{C}, 0)$ *with aligned singularities, where* $\mathcal{U}$ *is an open subset of* $\mathbb{C}^{n+1}$ *and* $\dim_{\mathbf{0}}\Sigma f_t = k$;

iii) for all $n \geqslant 3$, *for all k, the Zariski Multiplicity Conjecture is true for families of reduced analytic hypersurfaces* $f_t : (\mathcal{U}, \mathbf{0}) \to (\mathbb{C}, 0)$ *with aligned singularities, where* $\mathcal{U}$ *is an open subset of* $\mathbb{C}^{n+1}$ *and* $\dim_{\mathbf{0}}\Sigma f_t = k$.

Proof. Certainly, iii) implies ii). We will show that ii) implies i) and that i) implies iii).

Suppose that ii) is true for some $k \geqslant 1$. Let $f_t : (\mathcal{U}, \mathbf{0}) \to (\mathbb{C}, 0)$ be an analytic family, where $\mathcal{U}$ is an open subset of $\mathbb{C}^{n+1}$, $n \geqslant 3$, $\dim_{\mathbf{0}}\Sigma f_t = 0$, and such that the local ambient topological type of the hypersurfaces $V(f_t)$ at the origin is independent of t. Then clearly, the family

$$\tilde{f}_t : (\mathcal{U} \times \mathbb{C}^k, \mathbf{0}) \to (\mathbb{C}, 0)$$

defined by $\tilde{f}_t(\mathbf{z}, \mathbf{w}) := f_t(\mathbf{z})$ is a family of aligned singularities of dimension k with constant topological type.

Hence, by ii), $\mathrm{mult}_{\mathbf{0}}\tilde{f}_t$ is independent of t. Now, as $\mathrm{mult}_{\mathbf{0}}f_t$ clearly equals $\mathrm{mult}_{\mathbf{0}}\tilde{f}_t$, we are finished with the implication that ii) implies i).

The interesting implication is, of course, that i) implies iii). Ideally, we would like to be able to select linear coordinates, $\mathbf{z}$, for $\mathbb{C}^{n+1}$ which are aligning for f_t at the origin for all small t; however, a proof that this is possible seems problematic. We will avoid needing such a result by being somewhat devious and applying the Baire Category Theorem.

Suppose that i) is true, and that we have a family of reduced analytic hypersurfaces $f_t : (\mathcal{U}, \mathbf{0}) \to (\mathbb{C}, 0)$ with aligned singularities, where $\mathcal{U}$ is an open subset of $\mathbb{C}^{n+1}$, $n \geqslant 3$, $\dim_{\mathbf{0}}\Sigma f_t = k$, and such that the local ambient topological type of the hypersurfaces $V(f_t)$ at the origin is independent of t.

Let t_m be an infinite sequence in $\mathbb{C}$ which approaches 0, e.g. $t_m = \frac{1}{m}$. For each t_m, there exists a generic subset of $PGL(\mathbb{C}^{n+1})$ representing aligned coordinates for f_{t_m}. We may apply the Baire Category Theorem to conclude that there exists a choice of coordinates, $\mathbf{z}$, which is aligning for f_0 and for f_{t_m} for all m. Let us fix such a choice of coordinates.

Then, by 7.8, the Lê numbers $\lambda^i_{f_0,\mathbf{z}}(\mathbf{0})$ are equal to the Lê numbers $\lambda^i_{f_{t_m},\mathbf{z}}(\mathbf{0})$ for all large m. By an inductive application of 4.5, if we take $0 \ll j_0 \ll j_1 \ll \cdots \ll j_{s-1}$, then $f_0 + z_0^{j_0} + z_1^{j_1} + \cdots + z_{s-1}^{j_{s-1}}$ has an isolated singularity at the origin; this implies that, for all small t, $f_t + z_0^{j_0} + z_1^{j_1} + \cdots + z_{s-1}^{j_{s-1}}$ has, at worst, an isolated singularity at the origin. Moreover, 4.5 tells us that $f_0 + z_0^{j_0} + z_1^{j_1} + \cdots + z_{k-1}^{j_{k-1}}$ has the same Milnor number at the origin as $f_{t_m} + z_0^{j_0} + z_1^{j_1} + \cdots + z_{k-1}^{j_{k-1}}$ for all large m. As the Milnor number at the origin in the family $f_t + z_0^{j_0} + z_1^{j_1} + \cdots + z_{k-1}^{j_{k-1}}$ is upper-semicontinuous, it follows that, in fact, the Milnor number in this family is independent of t for all small t.

Hence, by **[L-R]**, the local, ambient topological type is independent of t in the family $f_t + z_0^{j_0} + z_1^{j_1} + \cdots + z_{k-1}^{j_{k-1}}$. Since we are assuming i), this implies that the multiplicity is independent of t in the family $f_t + z_0^{j_0} + z_1^{j_1} + \cdots + z_{k-1}^{j_{k-1}}$. Finally, as the j_m's are arbitrarily large, this implies that the multiplicity is independent of t in the family f_t. □

Chapter 8. SUSPENDING SINGULARITIES

In [**Ok**], [**Sak**], and [**Se-Th**], the general question is addressed of how the structure of the Milnor fibre of $f(\mathbf{z}) + g(\mathbf{w})$, where $\mathbf{z}$ and $\mathbf{w}$ are disjoint sets of variables, depends on the Milnor fibres of f and g. However, in each of these papers, f and g have isolated singularities or are quasi-homogeneous. Sakamoto remarks at the end of his paper that, by using Lê's notion of a good stratification, he can prove his main lemmas without the isolated singularity assumptions. As we crucially need this result in the special case of $f(\mathbf{z}) + w^j$, we will use our results in the appendix to indicate how one needs to modify Sakamoto's proof.

After we describe the homotopy-type of the Milnor fibre of $f(\mathbf{z}) + w^j$, we will use this description to give a new generalization of the formula of Lê and Iomdine [**Lê2**] – a different generalization than the formulas of Chapter 4.

Proposition 8.1. *If $j \geqslant 2$, then up to homotopy, the Milnor fibre of $\tilde{h}(w, \mathbf{z}) := h(\mathbf{z}) + w^j$ at the origin is the one-point union (wedge) of $j-1$ copies of the suspension of the Milnor fibre of h at the origin.*

Proof. By Proposition A.14, we may use neighborhoods of the form $\mathbb{D}_\omega \times B_\epsilon$, $0 < \omega \ll \epsilon$, to define the Milnor fibre of $\tilde{h}$ at the origin.

Now, for $0 < |\xi| \ll \omega \ll \epsilon$, consider the map

$$(\mathbb{D}_\omega \times B_\epsilon) \cap V(h + w^j - \xi) \xrightarrow{\ w\ } \mathbb{D}_\omega.$$

This map is a proper, stratified submersion above all points of $\mathbb{D}_\omega - V(w^j - \xi)$, i.e. except at the j roots of ξ. Thus, except above these j points, the fibre is the same as that above 0, which is clearly nothing more than the Milnor fibre of h at the origin. In addition, above each point of $V(w^j - \xi)$, the fibre is $B_\epsilon \cap V(h)$, which is contractible.

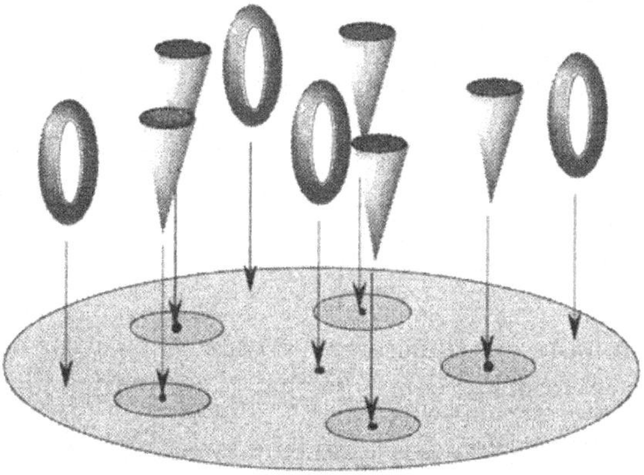

Figure 8.2. The fibres over the disc

In fact, around each point $\alpha_1, \ldots, \alpha_j$ in $V(w^j - \xi) \subseteq \mathbb{D}_\omega$, there is an arbitrarily small disc $\mathbb{D}_{\alpha_i} \subseteq \mathbb{D}_\omega$ above which the total space is contractible. We choose the $\mathbb{D}_{\alpha_i}$ disjoint. Connect all of the $\mathbb{D}_{\alpha_i}$ to the origin by disjoint paths. Let P denote the subset of $\mathbb{D}_\omega$ consisting of the paths; so, P is a contractible set which has exactly one point in common with each of the $\mathbb{D}_{\alpha_i}$, and the fibre above each point of P has the homotopy-type of the Milnor fibre of h at the origin.

The result now follows easily. For the details, we refer the reader to Sakamoto **[Sak]** – the remainder of our proof now follows his exactly. □

We shall now use 8.1 to give our second generalization of the formula of Lê and Iomdine.

Let $\mathcal{U}$ be an open neighborhood of the origin in $\mathbb{C}^{n+1}$, let $h : (\mathcal{U}, \mathbf{0}) \to (\mathbb{C}, 0)$ be an analytic function, and suppose that the linear from $L : \mathbb{C}^{n+1} \to \mathbb{C}$ is prepolar with respect to h at the origin.

The formula of Lê and Iomdine says that, if $\dim_{\mathbf{0}}\Sigma h = 1$, then, for all large j, $h + L^j$ has an isolated singularity at the origin and

$$b_n(h + L^j) \;=\; \mu(h + L^j) \;=\; b_n(h) - b_{n-1}(h) + j\sum_{\nu} m_\nu \delta_\nu(h),$$

where $b_i()$ denotes the i-th Betti number of the Milnor fibre of a function at the origin, μ denotes the Milnor number of the isolated singularity at the origin, the summation is over all components, ν, of Σh, m_ν is the local degree of L restricted to ν at the origin, and $\delta_\nu(h)$ is the Milnor number of a generic hyperplane slice of h at a point $\mathbf{p} \in \nu - \mathbf{0}$ sufficiently close to the origin.

This formula has, at least, two possible generalizations. One generalization is in terms of Lê numbers, as given in Chapter 4. But, while there are Morse inequalities between the Lê numbers and the Betti numbers of the Milnor fibre, the Lê numbers are not themselves (generally) Betti numbers of the Milnor fibre. So, one might ask for a generalization of the formula of Lê and Iomdine which generalizes the Betti number information.

In remainder of this chapter, we prove that, if $\dim_{\mathbf{0}}\Sigma h = s \geqslant 1$, then, for all large j, $\dim_{\mathbf{0}}\Sigma(h + L^j) = s - 1$ and

$$b_n(h + L^j) \;=\; b_n(h) - b_{n-1}(h) + j\big(b_{n-1}(h_{|_{V(L)}}) - \gamma^1_{h,L}(\mathbf{0})\big).$$

In the case where h has a one-dimensional critical locus at the origin, it is easy to show that this new formula reduces to that of Lê and Iomdine.

We consider this new Lê-Iomdine formula interesting because it implies that

$$b_{n-1}(h_{|_{V(L)}}) \geqslant \gamma^1_{h,L}(\mathbf{0}).$$

In terms of deformations, this says that the top possible non-zero Betti number of the Milnor fibre of $h_{|V(L)}$ is greater than or equal to $\gamma^1_{h,L}(\mathbf{0})$ for all h which have $V(L)$ as a prepolar slice. Thus, if we define h to be a *prepolar deformation* of $h_{|V(L)}$ precisely when $V(L)$ is a prepolar slice of h, we obtain a class of deformations of $h_{|V(L)}$ which give lower bounds on the top Betti number of the Milnor fibre. This also suggests that it might be useful to study prepolar deformations for which $\gamma^1_{h,L}(\mathbf{0})$ obtains its maximum value.

We will need

Proposition 8.3. *If $j \geqslant 2$ and $\mathcal{S}$ is a good stratification for h at the origin, then*

$$\{V(h + w^j) - \Sigma(h + w^j)\} \cup \{0 \times S \mid S \text{ is a singular stratum of } \mathcal{S}\}$$

is a good stratification for $h + w^j$ at the origin.

Proof. As $j \geqslant 2$, $\Sigma(h + w^j) = \{0\} \times \Sigma h$.

Let $\mathbf{p} = (0, \mathbf{q}) \in \{0\} \times \Sigma h$, where $S \in \mathcal{S}$, and let $\mathbf{p}_i = (u_i, \mathbf{q}_i)$ be a sequence of points in $\mathbb{C} \times \mathcal{U} - \{0\} \times \Sigma h$ such that $\mathbf{p}_i \to \mathbf{p}$ and

$$T_{\mathbf{p}_i} V(h + w^j - (h + w^j)_{|\mathbf{p}_i}) \to \mathcal{T}.$$

We wish to show that $T_{\mathbf{p}}(\{0\} \times S) = \{0\} \times T_{\mathbf{q}} S \subseteq \mathcal{T}$.

If $\mathcal{T} = T_{\mathbf{p}} V(w) = \{0\} \times \mathbb{C}^{n+1}$, then we are finished. So, suppose otherwise. Then, by taking a subsequence, we may assume that $\mathbf{q}_i \notin \Sigma h$ and that

$$T_{\mathbf{q}_i} V(h - h(\mathbf{q}_i)) \to \eta.$$

As $\mathcal{T}$ transversely intersects $T_{\mathbf{p}} V(w)$, $T_{\mathbf{p}_i} V(h + w^j - (h + w^j)_{|\mathbf{p}_i})$ transversely intersects $T_{\mathbf{p}_i} V(w - w_i)$ for all $\mathbf{p}_i$ close to $\mathbf{p}$. Thus,

$$\mathcal{T} \cap (\{0\} \times \mathbb{C}^{n+1}) = \lim T_{\mathbf{p}_i} V(h + w^j - (h + w^j)_{|\mathbf{p}_i}) \cap T_{\mathbf{p}_i} V(w - w_i) =$$

$$\lim T_{\mathbf{p}_i} V(h + w^j - (h + w^j)_{|\mathbf{p}_i}, w - w_i) = \lim T_{\mathbf{p}_i} V(h - h(\mathbf{q}_i), w - w_i) = \{0\} \times \eta.$$

Now, as S is a good stratum for h, $T_{\mathbf{q}}(S) \subseteq \eta$ and the proposition follows. □

Corollary 8.4. *If $V(z_0)$ is a prepolar slice for h at $\mathbf{0}$ then, for all $j \geqslant 2 + \lambda^0_{h,\mathbf{z}}(\mathbf{0})$, $V(z_0 - w)$ is a prepolar slice for $h + w^j$ at $\mathbf{0}$.*

Proof. In light of the proposition, all that we must show is that, for all $j \geqslant 2 + \lambda^0_{h,\mathbf{z}}(\mathbf{0})$, $\Sigma(h + w^j) \cap V(z_0 - w) = \Sigma(h + w^j{}_{|V(z_0 - w)})$.

Now,

$$\Sigma(h+w^j)\cap V(z_0-w)=(\{0\}\times\Sigma h)\cap V(z_0-w)=\{0\}\times(\Sigma h\cap V(z_0)).$$

On the other hand,

$$\Sigma(h+w^j{}_{|V(z_0-w)})=(\mathbb{C}\times\Sigma(h+z_0^j))\cap V(z_0-w).$$

But, near the origin and for $j\geqslant 2+\lambda^0_{h,\mathbf{z}}(\mathbf{0})$, $\Sigma(h+z_0^j)=\Sigma h\cap V(z_0)$ by 4.3.iii. The conclusion follows. □

Theorem 8.5. *If $V(z_0)$ is a prepolar slice of h at $\mathbf{0}$ then, for all $j\geqslant 2+\lambda^0_{h,\mathbf{z}}(\mathbf{0})$,*

$$b_n(h+z_0^j) \;=\; b_n(h)-b_{n-1}(h)+j\big(b_{n-1}(h_{|V(z_0)})-\gamma^1_{h,z_0}(\mathbf{0})\big),$$

where $b_i()$ denotes the i-th Betti number of the Milnor fibre at the origin.
In particular, $b_{n-1}(h_{|V(z_0)})\geqslant\gamma^1_{h,z_0}(\mathbf{0})$.

Proof.
After applying Proposition 3.1 to $h+w^j$ and the slice $V(z_0-w)$, and considering the long exact sequence of the pair, we have

$$b_{n+1}(h+w^j)-b_n(h+w^j)+b_n(h+z_0^j)=\left(\Gamma^1_{h+w^j,z_0-w}\cdot V(h+w^j)\right)_{\mathbf{0}}$$

which, by 4.3.v, equals $j\lambda^0_{h,\mathbf{z}}(\mathbf{0})$.
Now, as the Milnor fibre of $h+w^j$ has the homotopy-type of the one-point union of $j-1$ copies of the suspension of the Milnor fibre of h, we obtain

$$(j-1)b_n(h)-(j-1)b_{n-1}(h)+b_n(h+z_0^j)=j\lambda^0_{h,\mathbf{z}}(\mathbf{0}).$$

Using 3.1 on h itself and rearranging, we get

$$b_n(h+z_0^j)=b_n(h)-b_{n-1}(h)+j\left[\lambda^0_{h,\mathbf{z}}(\mathbf{0})-\left(\left(\Gamma^1_{f,z_0}\cdot V(f)\right)_{\mathbf{0}}-b_{n-1}(h_{|V(z_0)})\right)\right].$$

Finally, using the formula of Proposition 1.23 that

$$\left(\Gamma^1_{h,z_0}\cdot V(h)\right)_{\mathbf{0}}=\gamma^1_{h,z_0}(\mathbf{0})+\lambda^0_{h,z_0}(\mathbf{0}),$$

we obtain the desired result. □

The result of Theorem 8.5 is best thought of in terms of prepolar deformations: every prepolar deformation, h, of a fixed h_0 yields a lower bound on the top Betti number of the Milnor fibre of h_0.

One might hope that, by considering a prepolar deformation, h, for which $\gamma^1_{h,\mathbf{z}}(\mathbf{0})$ obtains its maximal value, one would actually obtain the top Betti number of the Milnor fibre of h_0. This seems unlikely however; certain singularities seem to be "rigid" with respect to prepolar deformations, in the weak sense that any prepolar deformation, h, has no polar curve at the origin.

Nonetheless, the lower bounds provided by prepolar deformations give new data which helps describe the Milnor fibre of a completely general affine hypersurface singularity; this data does not appear to follow from our Morse inequalities between the Betti numbers of the Milnor fibre and the Lê numbers of the hypersurface, as given in Theorem 3.3. As part of these Morse inequalities, we showed that $\lambda^0_{h_0,\tilde{z}}(\mathbf{0})$, provides an upper-bound on the top Betti number of the Milnor fibre of h_0. Also, it follows from 1.21 that if h is a prepolar deformation of h_0, then

$$\lambda^0_{h_0,\tilde{z}}(\mathbf{0}) = \lambda^1_{h,\mathbf{z}}(\mathbf{0}) + \gamma^1_{h,\mathbf{z}}(\mathbf{0}).$$

Thus, given a prepolar deformation, h, of h_0, we have bounded the top Betti number of the Milnor fibre of h_0:

$$\gamma^1_{h,\mathbf{z}}(\mathbf{0}) \leqslant\ b_{n-1}(h_0)\ \leqslant \lambda^1_{h,\mathbf{z}}(\mathbf{0}) + \gamma^1_{h,\mathbf{z}}(\mathbf{0}).$$

As $\lambda^0_{h_0,\tilde{z}}(\mathbf{0})$ is fixed, a prepolar deformation, h, with maximal $\gamma^1_{h,\mathbf{z}}(\mathbf{0})$ will have minimal $\lambda^1_{h,\mathbf{z}}(\mathbf{0})$. We prefer to call such a deformation a *minimal prepolar deformation*, instead of a maximal one. Note that a minimal prepolar deformation will not only have the maximal possible lower bound on the top Betti number of the Milnor fibre, it also provides the smallest difference between our general upper and lower bounds. One might hope that it is always possible to find a prepolar deformation, h, for which $\lambda^1_{h,\mathbf{z}}(\mathbf{0}) = 0$, for then we would have $b_{n-1}(h_0) = \gamma^1_{h,\mathbf{z}_0}(\mathbf{0})$; unfortunately, Proposition 1.31 implies that it is usually impossible to find such a deformation.

Chapter 9. CONSTANCY OF THE MILNOR FIBRATIONS

In this chapter, we prove what is perhaps our most important result, and certainly the result which requires the most machinery – we generalize the result of Lê and Ramanujam [**L-R**] as stated in Theorem 0.10 in the introduction. Basically, we prove that if the Lê numbers are constant in a one-parameter family, then the Milnor fibrations are constant in the family, **regardless of the dimension of the critical loci**.

Unfortunately, we do not obtain the result that the local, ambient topological-type of the hypersurfaces remains constant in the family. It is an open question whether the constancy of the Lê numbers is strong enough to imply this topological constancy.

While the idea behind our proof of this generalized Lê-Ramanujam is simple, the technical details are horrendous. It is this chapter alone which is responsible for the appendix of this book; we have relegated most of the technical details to the appendix. Before we prove the main result, there remain only two lemmas which we need (besides the results which appear in the appendix). Also, we will restate one of the results from the appendix in a form which is comprehensible without reading the entire appendix.

First, however, we wish to sketch the proof of the main theorem, so that the reader can see that the idea really is fairly easy. On the other hand, the proof is not straightforward - instead, it uses a trick which gives one very little insight as to why the result should be true.

Throughout this chapter, $\mathcal{U}$ will denote an open neighborhood of the origin in $\mathbb{C}^{n+1}$ and $f_t : (\mathcal{U}, \mathbf{0}) \rightarrow (\mathbb{C}, 0)$ will be an analytic family in the variables $\mathbf{z} = (z_0, \ldots, z_n)$. Let $s = \dim_{\mathbf{0}} \Sigma f_0$.

A sketch of the proof is as follows:

The result of Proposition 8.1 is that the Milnor fibre of $f_t + w^j$ at the origin is homotopy-equivalent to the one-point union of $j-1$ copies of the suspension of the Milnor fibre of f_t at the origin. So, it certainly seems reasonable to expect that the Milnor fibrations are independent of t in the family f_t if and only if the Milnor fibrations are independent of t in the family $f_t + w^j$. But why should the family $f_t + w^j$ be any easier to study than the family f_t itself?

It is easier because we have very nice hyperplanes defined by $L = w - z_0$ such that, when we take the sections $(f_t + w^j)_{|V(L)}$, we get the family $f_t + z_0^j$ which, for generic z_0 and for large j, is a family of singularities of one less dimension (by the results of Chapter 4); that is, $\dim_{\mathbf{0}} \Sigma(f_0 + z_0^j) = s - 1$. Moreover, Lê's attaching result (Theorem 0.9) tells us how the Milnor fibre of $f_t + w^j$ is obtained from the Milnor fibre of a generic hyperplane section. The Milnor fibre of $f_t + w^j$ is obtained from the Milnor fibre of $f_t + z_0^j$ by attaching $\left(\Gamma^1_{f_t + w^j, w - z_0} \cdot V(f_t + w^j)\right)_{\mathbf{0}}$ $(n+1)$-handles.

By induction on s, we may require the Milnor fibrations of $f_t + z_0^j$ to be independent of t. If we also require the number of attached $(n+1)$-handles to be independent of t, it seems reasonable to expect that the Milnor fibrations of the family $f_t + w^j$ should be independent of t and, thus, that the Milnor fibrations of f_t are independent of t.

The Lê numbers enter the picture because Lemma 4.3 says that, for large j, $\left(\Gamma^1_{f_t+w^j, w-z_0} \cdot V(f_t + w^j)\right)_{\mathbf{0}} = j\lambda^0_{f_t,\mathbf{z}}(\mathbf{0})$. Combining this with the Lê-Iomdine formulas of Theorem 4.5, we find that the inductive requirement that the Milnor fibrations of $f_t + z_0^j$ are independent of t amounts to requiring all the Lê numbers of f_t to be independent of t.

We first wish to prove a result which will tell us that the main theorem of this chapter is not vacuously true.

Lemma 9.1. *For all i with $0 \leqslant i \leqslant n$, for a generic linear choice of the coordinates $(z_0, \ldots, z_i)$, $(z_0, \ldots, z_i)$ is prepolar at the origin for f_t for all small t.*

Proof. Fix a good stratification, $\mathfrak{G}$, for f in a neighborhood, $\mathcal{V}$, of the origin. By refining if necessary, we may also assume that $\mathfrak{G}$ satisfies Whitney's condition a). We will also assume that

$$S := \mathcal{V} \cap \{t\text{-axis}\} - \{\mathbf{0}\} = \mathcal{V} \cap (\mathbb{C} \times \{\mathbf{0}\}) - \{\mathbf{0}\}$$

and $\{\mathbf{0}\}$ are strata of $\mathfrak{G}$. As the function t has isolated stratified critical values, $V(t - t_0)$ transversely intersects all strata of $\mathfrak{G}$ near $(t_0, \mathbf{0})$; hence, $V(t - t_0) \cap \mathfrak{G}$ provides a good stratification for f_{t_0} at $\mathbf{0}$ which still satisfies Whitney's condition a).

By induction on i, we will prove that: for all i with $0 \leqslant i \leqslant n$, for a generic linear choice of the coordinates $(z_0, \ldots, z_i)$, $(z_0, \ldots, z_i)$ is prepolar at the origin for f_{t_0} with respect to the good stratification $V(t - t_0) \cap \mathfrak{G}$ for all small, non-zero t_0.

$i = 0$: Using the terminology of Goresky and MacPherson [**G-M2**], the set of degenerate conormal covectors to S is a complex analytic subvariety of codimension $\geqslant 1$ inside the total space of the conormal bundle of S inside $\mathbb{C}^{n+2}$ (see [**G-M2**, Prop. 1.8, p.44]). Projectivizing and dualizing, this says that the set Ω, defined by

$$\{(\mathbf{p}, H) \in S \times G_n(\mathbb{C}^{n+1}) \mid T_{\mathbf{p}}(\mathbb{C} \times H) \supseteq \text{a generalized tangent plane of } \mathfrak{G} \text{ at } \mathbf{p}\},$$

has dimension at most n. Hence, $\overline{\Omega} \subseteq (\mathcal{V} \cap (\mathbb{C} \times \mathbf{0})) \times G_n(\mathbb{C}^{n+1})$ has dimension at most n and the fibre over $\mathbf{0}$, call it Ψ, must therefore have dimension at most

$n-1$. Thus, $G_n(\mathbb{C}^{n+1}) - \Psi$ is open and dense in $G_n(\mathbb{C}^{n+1})$. We claim that, for all H in $G_n(\mathbb{C}^{n+1}) - \Psi$, H is a prepolar slice for f_t at $\mathbf{0}$ for all small, non-zero t.

Certainly, if $H \notin \Psi$, then, for all small, non-zero t_0, $T_{(t_0,\mathbf{0})}(\mathbb{C} \times H)$ contains no generalized tangent plane from $\mathfrak{G}$ at $(t_0, \mathbf{0})$. Also, Whitney's condition a) guarantees that all limiting tangent planes from strata of $\mathfrak{G}$ at $(t_0, \mathbf{0})$ actually contains $T_{(t_0,\mathbf{0})}S = \mathbb{C} \times \mathbf{0}$. Combining these two facts, it follows easily that $T_{\mathbf{0}}H$ contains no generalized tangent plane from $V(t-t_0) \cap \mathfrak{G}$ at $\mathbf{0}$. Thus, H is prepolar for f_{t_0} at $\mathbf{0}$ with respect to the good stratification $V(t-t_0) \cap \mathfrak{G}$. (Actually, this implies much more – it implies that H is *polar*, as defined in [**Mas3**].)

$i \geqslant 1$: Now, assume that we have already chosen $(z_0, \dots, z_{i-1})$ generically so that $(z_0, \dots, z_{i-1})$ is prepolar at the origin for f_{t_0} with respect to the good stratification $V(t-t_0) \cap \mathfrak{G}$ for all small, non-zero t_0.

Then, there exists a good stratification, $\mathfrak{G}'$, for $f_{|V(z_0,\dots,z_{i-1})}$ at the origin which satisfies Whitney's condition a). Though $\mathfrak{G}'$ may not necessarily be chosen to equal $\mathfrak{G} \cap V(z_0, \dots, z_{i-1})$, after refining $\mathfrak{G}'$ using Proposition A.2, we may certainly assume that each stratum of $\mathfrak{G}'$ is contained in a stratum of $\mathfrak{G}$ and that, in some neighborhood of the origin, $(\mathbb{C} \times \mathbf{0}) - \mathbf{0}$ and $\mathbf{0}$ are strata of $\mathfrak{G}'$.

By the $i = 0$ case, for a generic choice of z_i, $V(z_i)$ is a prepolar slice for $f_{t_0|V(z_0,\dots,z_{i-1})}$ at the origin with respect to $V(t-t_0) \cap \mathfrak{G}'$ for all small non-zero t_0. But, as each stratum of $\mathfrak{G}'$ is contained in a stratum of $\mathfrak{G}$, this last statement is stronger than saying that $V(z_i)$ is prepolar with respect to $V(t-t_0) \cap \mathfrak{G}$. This concludes the induction.

Finally, to finish the proof, one simply chooses coordinates as generic as given above and generically enough so that the coordinates are prepolar for f_0 at the origin. □

We shall also need the following uniform version of Proposition 4.19:

Proposition 9.2. *If $(z_0, \dots, z_i)$ is prepolar at the origin for f_t for all small t, then, for all large j, $(z_1, \dots, z_i)$ is prepolar for $f_t + z_0^j$ for all small t.*

Proof. In light of Corollary 4.18, what we need to show is that, for all large j, for all k with $0 \leqslant k \leqslant i$ and for all small t_0,

$$(*) \qquad \dim_{\mathbf{0}} \Gamma^{k+1}_{f_{t_0},\mathbf{z}} \cap V\left(t - t_0, \frac{\partial f}{\partial z_0} + j z_0^{j-1}\right) \cap V(z_1, \dots, z_k) \leqslant 0.$$

In fact, we only have to show that $(*)$ holds for small **non-zero** t_0, for then – by 4.19 – we may impose the extra largeness condition on j so that $(*)$ also holds for $t_0 = 0$.

By Lemma 4.14, for all small non-zero t_0, $\Gamma^{k+1}_{f_{t_0},\mathbf{z}} = \Gamma^{k+2}_{f,(t,\mathbf{z})} \cap V(t-t_0)$ as sets, in a neighborhood of the origin. In addition, by Theorem 1.28, $\gamma^{k+1}_{f_{t_0},\mathbf{z}}(\mathbf{0})$ exists for all small t_0; thus, $\dim_{\mathbf{0}}\Gamma^{k+1}_{f_{t_0},\mathbf{z}} \cap V(z_0, z_1, \ldots, z_k) \leqslant 0$. Putting these two facts together, we find that, for all small non-zero t_0,

$$\dim_{\mathbf{0}} V(t - t_0) \cap \Gamma^{k+2}_{f,(t,\mathbf{z})} \cap V(z_0, z_1, \ldots, z_k) \leqslant 0.$$

Let W denote the union of those irreducible analytic components, C, of $\Gamma^{k+2}_{f,(t,\mathbf{z})} \cap V(z_1, \ldots, z_k)$, at the origin, which satisfy the property that the t-axis is contained in $C \cap V(z_0)$. By the above, $W \cap V(z_0)$ is at most one-dimensional at the origin and, for all small non-zero t_0, $\Gamma^{k+1}_{f_{t_0},\mathbf{z}} \cap V(z_1, \ldots, z_k) = W \cap V(t - t_0)$ as germs of sets at $(t_0, \mathbf{0})$.

It follows that each irreducible component of W at the origin is at most 2-dimensional. We would like to show that, for all large j, $W \cap V\left(\frac{\partial f}{\partial z_0} + jz_0^{j-1}\right)$ is at most 1-dimensional at the origin, for then – by intersecting with $V(t - t_0)$ – we obtain $(*)$.

But, this is easy. For if a 2-dimensional component C of W is contained in $V\left(\frac{\partial f}{\partial z_0} + j_0 z_0^{j_0-1}\right)$ and $V\left(\frac{\partial f}{\partial z_0} + j_1 z_0^{j_1-1}\right)$ for $j_0 \neq j_1$, then $C \subseteq V(z_0)$. However, we know that $W \cap V(z_0)$ is at most 1-dimensional. Hence, there are only a finite number of "bad" values for j. □

Proposition 9.3 *Suppose that $V(z_0)$ is a prepolar slice for f_t at the origin for all small t, and that $V(t)$ does not occur as the limit of tangent spaces to level hypersurfaces of f or $f_{|V(z_0)}$ at the origin. Then, for all small non-zero t_0, there is a natural inclusion of pairs of Milnor fibres $(F_{f_{t_0},\mathbf{0}}, F_{f_{t_0},\mathbf{0}} \cap V(z_0)) \hookrightarrow (F_{f_0,\mathbf{0}}, F_{f_0,\mathbf{0}} \cap V(z_0))$.*

If we also assume that the intersection number $\left(\Gamma^1_{f_t,z_0} \cdot V(f_t)\right)_{\mathbf{0}}$ is independent of t for all small t, then we have the following three results:

i) if the inclusion $F_{f_{t_0},\mathbf{0}} \cap V(z_0) \hookrightarrow F_{f_0,\mathbf{0}} \cap V(z_0)$ induces isomorphisms on integral homology groups, then so does the inclusion $F_{f_{t_0},\mathbf{0}} \hookrightarrow F_{f_0,\mathbf{0}}$;

ii) if $s \leqslant n-2$ and the inclusion $F_{f_{t_0},\mathbf{0}} \cap V(z_0) \hookrightarrow F_{f_0,\mathbf{0}} \cap V(z_0)$ is a homotopy-equivalence, then so is the inclusion $F_{f_{t_0},\mathbf{0}} \hookrightarrow F_{f_0,\mathbf{0}}$, and the fibre homotopy-type of the Milnor fibrations is independent of t for all small t; and

iii) if $s \leqslant n-3$ and the inclusion $F_{f_{t_0},\mathbf{0}} \cap V(z_0) \hookrightarrow F_{f_0,\mathbf{0}} \cap V(z_0)$ is a homotopy-equivalence, then the inclusion $F_{f_{t_0},\mathbf{0}} \hookrightarrow F_{f_0,\mathbf{0}}$ is a diffeomorphism and, moreover, the diffeomorphism-type of the Milnor fibrations is independent of t for all small t.

Proof. This is primarily a restatement of Theorem A.13 from the appendix.

The condition on $V(t)$ is that $V(t)$ is not in the Thom set of either f or $f_{|V(z_0)}$ at the origin (see Definition A.8). Therefore, by Proposition A.9, the families f_t and $f_{t|V(z_0)}$ satisfy the universal conormal condition at the origin (see Definition

A.10) and this produces the inclusion of pairs of Milnor fibres. Now, i) and ii) follow from Theorem A.13, and iii) follows from ii) together with Proposition A.12. □

Finally, we are able to prove the main result of this chapter: our generalization of part of the result of Lê and Ramanujam. Essentially, we prove that the constancy of the Lê numbers in a family implies the constancy of the Milnor fibrations in the family.

Theorem 9.4. *Let $s := \dim_{\mathbf{0}}\Sigma f_0$. Suppose that, for all t small, $(z_0, \ldots, z_{s-1})$ is prepolar for f_t at $\mathbf{0}$ and that the Lê numbers, $\lambda^i_{f_t,\mathbf{z}}(\mathbf{0})$, are independent of t for each i with $0 \leqslant i \leqslant s$. Then,*

i) the homology of the Milnor fibre of f_t at the origin is independent of t for all t small;

if $s \leqslant n-2$,

ii) the fibre homotopy-type of the Milnor fibrations of f_t at the origin is independent of t for all t small;

and, if $s \leqslant n-3$,

iii) the diffeomorphism-type of the Milnor fibrations of f_t at the origin is independent of t for all t small.

Proof. By induction on s.

For $s = 0$, this is the result of Lê and Ramanujam [**L-R**]. Now, assume that $s \geqslant 1$ and that we know the result for families of hypersurfaces with critical loci of dimension $\leqslant s-1$.

Let j be so large that the uniform Lê-Iomdine formulas of Theorem 4.15 hold and so large that Proposition 9.2 holds. Finally, using that $\lambda^0_{f_t,\mathbf{z}}(\mathbf{0})$ is independent of t, let j be so large that $j \geqslant 2 + \lambda^0_{f_t,\mathbf{z}}(\mathbf{0})$ for all small t, so that we may apply Lemma 4.3 and Corollary 8.4.

Consider the family $f_t + w^j$, where w is a variable disjoint from the z's. The dimension of the critical locus at $t = 0$ is still equal to s. As the Lê numbers of f_t are independent of t, we may apply Theorem 6.5 to conclude that $V(t)$ is not the limit of tangent spaces to level hypersurfaces of f at the origin. It follows trivially that $V(t)$ is not the limit of tangent spaces to level hypersurfaces of $f + w^j$ at the origin. Moreover, using the uniform Lê-Iomdines formulas of Theorem 4.15, we find that the Lê numbers of $(f_t+w^j)_{|V(z_0-w)} = f_t+z_0^j$ at $\mathbf{0}$ with respect to $(z_1, \ldots, z_{s-1})$ are independent of t, and that $\dim_{\mathbf{0}}\Sigma(f_0 + z_0^j) = s-1$.

Therefore, by our inductive hypothesis, we already have the constancy results for the family $(f_t + w^j)_{|V(z_0-w)}$. In addition, we may use Theorem 6.5 again to conclude that $V(t)$ is not the limit of tangent spaces to level hypersurfaces of $(f+w^j)_{|V(z_0-w)}$ at the origin. By 8.4, $V(z_0 - w)$ is a prepolar slice for $f_t + w^j$ at the origin for all small t. Also, by Lemma 4.3.v, $\left(\Gamma^1_{f_t+w^j, w-z_0} \cdot V(f_t + w^j)\right)_{\mathbf{0}} = j\lambda^0_{f_t,\mathbf{z}}(\mathbf{0})$, which is independent of t for all small t. Thus, we are in a position to apply Proposition 9.3 to the family $f_t + w^j$ and we conclude the desired constancy results for this family.

Finally, now that we know the results for $f_t + w^j$, we apply Proposition A.16 to conclude that the result actually holds for f_t itself. □

Remark 9.5. While we are not very fond of discussing it, for a fixed function h and a fixed point $\mathbf{p}$, there exist *generic Lê numbers* of h at $\mathbf{p}$ (see Theorem 10.19) – that is, as one varies the linear choice of coordinates, $\mathbf{z}$, through coordinates for which the Lê numbers are defined, one finds a generic value for each of the Lê numbers, $\lambda^i_{h,\mathbf{z}}(\mathbf{p})$; let us denote this generic value simply by $\lambda^i_h(\mathbf{p})$. We know that such generic values exist and are analytic invariants because of the relation between the Lê numbers and polar multiplicities, as described later in Chapter 10.

We are not fond of discussing these generic Lê numbers because the Lê numbers are intended to be effectively calculable, and we know of no effective way of knowing when a coordinate choice is sufficiently generic to give $\lambda^i_h(\mathbf{p})$. We do know, by Corollary 4.16, that the tuple of generic Lê numbers $(\lambda^s_h(\mathbf{p}), \dots, \lambda^0_h(\mathbf{p}))$ is minimal with respect to the lexigraphic ordering.

We mention all this here because the tuple $(\lambda^s_h(\mathbf{p}), \dots, \lambda^0_h(\mathbf{p}))$ is an analytic invariant, and so the reader may wonder whether Theorem 9.4 stills holds under the assumption that the generic Lê numbers of f_t are independent of t at the origin. The answer to this question is easily seen to be: yes. This follows quickly from 9.4 itself.

Suppose that the generic Lê numbers of f_t at the origin are independent of t. Choose coordinates $\mathbf{z}$ so generic that $\mathbf{z}$ is prepolar for f_t at the origin for all small t and so that $\mathbf{z}$ is generic enough to give the generic Lê numbers of f_0 at the origin. Then, for all small t, we have

$$(\lambda^s_{f_t}(\mathbf{0}), \dots, \lambda^0_{f_t}(\mathbf{0})) \leqslant (\lambda^s_{f_t,\mathbf{z}}(\mathbf{0}), \dots, \lambda^0_{f_t,\mathbf{z}}(\mathbf{0}))$$

$$\leqslant (\lambda^s_{f_0,\mathbf{z}}(\mathbf{0}), \dots, \lambda^0_{f_0,\mathbf{z}}(\mathbf{0})) = (\lambda^s_{f_0}(\mathbf{0}), \dots, \lambda^0_{f_0}(\mathbf{0})),$$

where we are using the lexigraphic ordering. Since

$$(\lambda^s_{f_t}(\mathbf{0}), \dots, \lambda^0_{f_t}(\mathbf{0})) = (\lambda^s_{f_0}(\mathbf{0}), \dots, \lambda^0_{f_0}(\mathbf{0})),$$

it follows that $(\lambda^s_{f_t,\mathbf{z}}(\mathbf{0}), \dots, \lambda^0_{f_t,\mathbf{z}}(\mathbf{0}))$ equals the tuple at $t = 0$, and so we may apply Theorem 9.4.

Chapter 10. OTHER CHARACTERIZATIONS OF THE LÊ CYCLES

In this chapter, we give some alternative characterizations of the Lê cycles and Lê numbers of a hypersurface singularity. These alternative characterizations lead to a generalization of the Lê numbers which can be applied to any perverse sheaf. From this more general viewpoint of perverse sheaves, the case of the Lê numbers of a function f is just the case where the underlying perverse sheaf is the sheaf of vanishing cycles along f.

As a consequence of Theorem 3.3, the Lê cycles can be characterized formally by requiring that the alternating sum of the Lê numbers yields the reduced Euler characteristic of the Milnor fibre at each point - provided that we know that a fixed choice of coordinates is prepolar at **every** point. We will show this now.

Lemma 10.1. *Let X be an analytic subset containing the origin in some $\mathbb{C}^N$ and let $\{S_\alpha\}$ be an analytic stratification of X with connected strata. Let $p : \mathbb{C}^N \to \mathbb{C}^k$ be a linear map such that $p_{|S_\alpha}$ is a submersion if $\dim S_\alpha \geqslant k$ and $\dim_{\mathbf{0}}\left(p^{-1}(\mathbf{0}) \cap \overline{S}_\alpha\right) = 0$ if $\dim S_\alpha \leqslant k-1$.*

Then, for generic linear $\pi : \mathbb{C}^k \to \mathbb{C}^{k-1}$, there exists a refinement $\{R_\beta\}$ of $\{S_\alpha\}$ in some neighborhood of the origin which preserves the strata of dimension greater than or equal to k and such that $\pi \circ p_{|R_\beta}$ is a submersion if $\dim R_\beta \geqslant k-1$ and $\dim_{\mathbf{0}}\left((\pi\circ p)^{-1}(\mathbf{0}) \cap \overline{R}_\beta\right) = 0$ if $\dim R_\beta \leqslant k-2$.

Proof. Let X' denote the union of those strata S_α such that $\dim S_\alpha \leqslant k-1$. Then, as $\dim_{\mathbf{0}}(p^{-1}(\mathbf{0}) \cap X') = 0$, there exists an open neighborhood, $\mathcal{U}$, of the origin in $\mathbb{C}^N$ and an open neighborhood of the origin, $\mathcal{V}$, in $\mathbb{C}^k$ such that the restriction of p to a map from $\mathcal{U} \cap X'$ to $\mathcal{V}$ is finite. Therefore, as the conclusion of the lemma is purely local in nature, we may reduce ourselves to considering only the case where $p_{|X'}$ is a finite map.

Thus, $p(X')$ is an analytic subset of $\mathbb{C}^k$ of dimension at most $k-1$ and so for generic lines, L, through $\mathbf{0}$ in $\mathbb{C}^k$ we have that $L \cap p(X') = \{\mathbf{0}\}$ near the origin. Hence, for generic linear $\pi : \mathbb{C}^k \to \mathbb{C}^{k-1}$, we must have that $\pi^{-1}(\mathbf{0}) \cap p(X') = \{\mathbf{0}\}$, and so

$$(\pi\circ p)^{-1}(\mathbf{0}) \cap X' = p^{-1}\left(\pi^{-1}(\mathbf{0}) \cap p(X')\right) \cap X' = p^{-1}(\mathbf{0}) \cap X'.$$

But, by hypothesis, $p^{-1}(\mathbf{0}) \cap X' = \{\mathbf{0}\}$, and thus - for such a generic π - any refinement of X' will give a refinement of $\{S_\alpha\}$ which preserves the strata of dimension greater than or equal to k and realizes the desired intersection condition.

Now, we have the restriction $\pi \circ p : X' \to \mathbb{C}^{k-1}$. By the above, if we once again take sufficiently small neighborhoods around the origins, $\pi \circ p$ restricts to a finite map on X'. Assume then that $(\pi \circ p)_{|_{X'}}$ is finite. If $\dim X' = k-1$, then $\pi \circ p$ is a submersion on $X' - W$, where W is a closed subset of X' of dimension at most $k-2$. Thus, we may refine the stratification so that $X' \cap (\pi \circ p)^{-1}(\pi \circ p)(W)$ is a union of strata, where the strata have dimension at most $k-2$ since $\pi \circ p$ restricted to X' is a finite map. This yields the desired refinement $\{R_\beta\}$. □

Proposition 10.2. *Let $\mathcal{U}$ be an open subset of $\mathbb{C}^{n+1}$ containing the origin, and let $h : (\mathcal{U}, \mathbf{0}) \to (\mathbb{C}, 0)$ be an analytic map. Then, for a generic linear choice, $\mathbf{z}$, of coordinate systems for $\mathbb{C}^{n+1}$, there exists an open neighborhood $\mathcal{U}'$ such that $\mathbf{z}$ is prepolar for $h_{|_{\mathcal{U}'}}$, i.e. for all $\mathbf{p} \in \mathcal{U}' \cap V(h)$, if we let s denote $\dim_{\mathbf{p}} \Sigma h$, then $(z_0, \ldots, z_{s-1})$ is prepolar for h at $\mathbf{p}$.*

Proof. Begin by applying the lemma to $X = V(h)$, endowed with some good stratification, and $p = \mathrm{id} : \mathbb{C}^{n+1} \to \mathbb{C}^{n+1}$. By an inductive application of the lemma, for a generic linear choice of coordinates, $\mathbf{z}$, for $\mathbb{C}^{n+1}$, we arrive at a stratification $\{S_\alpha\}$, which is a refinement of the original good stratification (in a possibly smaller neighborhood of the origin), such that for each i, the map $(z_0, \ldots, z_i)$ is a submersion when restricted to a stratum of dimension greater than or equal to $i+1$. This would say precisely that $(z_0, \ldots, z_n)$ is prepolar at each point of $V(h)$ with respect to the stratification $\{S_\alpha\}$ - provided that $\{S_\alpha\}$ is actually a good stratification for h.

But, any refinement of a good stratification is another good stratification - except for the fact that we may have refined the smooth part of $V(h)$, which we required to be a stratum of any good stratification. However,

$$\{V(h) - \Sigma h\} \cup \{S_\alpha \mid S_\alpha \subseteq \Sigma h\}$$

is certainly a good stratification for h, and what remains for us to show is that, for generic coordinates, $\mathbf{z}$, for each $\mathbf{p} \in \Sigma h$, $V(z_0 - p_0, \ldots, z_{s-1} - p_{s-1})$ transversely intersects the smooth part of $V(h)$ near $\mathbf{p}$, where again $s = \dim_{\mathbf{p}} \Sigma h$.

Assume then that we have chosen $\mathbf{z}$ generic as above and also generic enough so that $(z_0, \ldots, z_{s-1})$ is prepolar for h at $\mathbf{0}$, where $s = \dim_{\mathbf{0}} \Sigma h$. By Theorem 1.28, $\gamma^s_{h,\mathbf{z}}(\mathbf{0})$ exists, and so $\gamma^s_{h,\mathbf{z}}(\mathbf{p})$ exists for all $\mathbf{p}$ near $\mathbf{0}$. But, by Remark 1.10, this implies that if $\mathbf{p} \in \Sigma h$, then

$$\Sigma(h_{|_{V(z_0 - p_0, \ldots, z_{k-1} - p_{k-1})}}) = V(z_0 - p_0, \ldots, z_{k-1} - p_{k-1}) \cap \Sigma h$$

at $\mathbf{p}$, i.e. $V(z_0 - p_0, \ldots, z_{s-1} - p_{s-1})$ transversely intersects the smooth part of $V(h)$ near $\mathbf{p}$ (and, of course, $s = \dim_{\mathbf{0}} \Sigma h \geqslant \dim_{\mathbf{p}} \Sigma h$ for all $\mathbf{p}$ near $\mathbf{0}$). □

Theorem 10.3. *For a generic linear choice of coordinates, $\mathbf{z}$, for $\mathbb{C}^{n+1}$, the Lê cycles are a collection of analytic cycle germs, $\Lambda^i_{h,\mathbf{z}}$, in Σh at the origin such that each $\Lambda^i_{h,\mathbf{z}}$ is purely i-dimensional and properly intersects $V(z_0, \ldots, z_{i-1})$ at* *the origin, and*

$$\tilde{\chi}(F_{h,\mathbf{p}}) = \sum_{i=0}^{s}(-1)^{n-i}\left(\Lambda^i_{h,\mathbf{z}} \cdot V(z_0 - p_0, \dots, z_{i-1} - p_{i-1})\right)_{\mathbf{p}},$$

for all $\mathbf{p} \in \Sigma h$ *near* $\mathbf{0}$, *where* $\tilde{\chi}(F_{h,\mathbf{p}})$ *is the reduced Euler characteristic of the Milnor fibre of* h *at* $\mathbf{p}$.

Moreover, if $\mathbf{z}$ *is any linear coordinate system such that such cycles exist, then they are unique.*

Proof. The first statement follows immediately from the previous proposition and Theorem 3.3. As for the uniqueness assertion, this is a fairly standard argument for constructible functions.

Suppose that we had two such collections, $\Lambda^i_{h,\mathbf{z}}$ and $\Omega^i_{h,\mathbf{z}}$. Let s denote $\dim_{\mathbf{0}}\Sigma h$. We will show that $\Lambda^i_{h,\mathbf{z}}$ and $\Omega^i_{h,\mathbf{z}}$ agree by downward induction on i.

For a generic point, $\mathbf{p}$, in an s-dimensional component, ν, of the support of $\Lambda^s_{h,\mathbf{z}}$, $\mathbf{p}$ will be a smooth point of ν, $V(z_0 - p_0, \dots, z_{s-1} - p_{s-1})$ will transversely intersect ν at $\mathbf{p}$, and $\mathbf{p}$ will not be in the support of any of the lower-dimensional $\Lambda^i_{h,\mathbf{z}}$ or $\Omega^i_{h,\mathbf{z}}$. Thus, at such a $\mathbf{p}$,

$$\left(\Omega^s_{h,\mathbf{z}} \cdot V(z_0 - p_0, \dots, z_{s-1} - p_{i-1})\right)_{\mathbf{p}} = \left(\Lambda^s_{h,\mathbf{z}} \cdot V(z_0 - p_0, \dots, z_{s-1} - p_{i-1})\right)_{\mathbf{p}}.$$

Of course, the same conclusion would have followed it we had chosen a generic point, $\mathbf{p}$, in an s-dimensional component, ν, of the support of $\Omega^s_{h,\mathbf{z}}$, It follows that $\Lambda^s_{h,\mathbf{z}} = \Omega^s_{h,\mathbf{z}}$.

Now, suppose that we have shown that $\Lambda^i_{h,\mathbf{z}} = \Omega^i_{h,\mathbf{z}}$ for all i greater than some k. Then,

$$\sum_{i=0}^{k}(-1)^{n-i}\left(\Lambda^i_{h,\mathbf{z}} \cdot V(z_0 - p_0, \dots, z_{i-1} - p_{i-1})\right)_{\mathbf{p}} =$$

$$\sum_{i=0}^{k}(-1)^{n-i}\left(\Omega^i_{h,\mathbf{z}} \cdot V(z_0 - p_0, \dots, z_{i-1} - p_{i-1})\right)_{\mathbf{p}},$$

and we repeat the argument above with k in place of s. The conclusion follows. □

Remark 10.4. Theorem 10.3 leaves us with a very strange set of affairs; it tells us that, for generic $\mathbf{z}$, we could have defined the Lê cycles as above. This means that the Lê cycles and hence, the Lê numbers, are determined by the choice of $\mathbf{z}$ and the data of the Euler characteristic of the Milnor fibre at each point. But, had we defined the Lê cycles and numbers this way, then we would have produced the Morse inequalities on the Betti numbers of the Milnor fibres in Theorem 3.3 from seemingly much less data.

These Morse inequalities would result from a chain complex of $\mathbb{C}$-vector spaces which has cohomology equal to the reduced cohomology of the Milnor fibre and in which the terms have dimensions given by the Lê numbers. But why is there such a chain complex?

While a complete explanation of this phenomenon is beyond the scope and methods of these notes, we denote the remainder of the chapter to, at least, a partial account. We will use techniques which are very different from those used up to this point. We give these results – most without proof – both as concluding remarks and to indicate the direction of our future research. These results appear in [**Mas5**].

The proper solution to our puzzle requires the language and machinery of the derived category, perverse sheaves, and nearby and vanishing cycles; and we will have to assume some of the basic results as described in [**BBD**], [**Br**], [**De**], [**G-M1**], [**K-S2**], and [**Lê2**].

Let X be an analytic germ of an s-dimensional space which is embedded in some affine space, $M := \mathbb{C}^{n+1}$, so that the origin is a point of X. We are interested in the germ of X at the origin.

$\mathbf{F}^\bullet$ will be used to denote a bounded, constructible sheaf on X or M. If $\mathbf{F}^\bullet$ is perverse, then its non-zero cohomology groups lie in non-positive dimensions. For our purposes, it will be more convenient to shift the complex so that the cohomology is in non-negative dimensions. Hence, we define $\mathbf{F}^\bullet$ on X to be *positively perverse* if and only if $\mathbf{F}^\bullet[s]$ is perverse on X. For instance, the constant sheaf $\mathbb{C}^\bullet_M$ is positively perverse on M.

If $f : X \to \mathbb{C}$ is an analytic map and $\mathbf{F}^\bullet$ is a constructible complex on X, then we denote the sheaves of nearby and vanishing cycles of $\mathbf{F}^\bullet$ along f by $\psi_f\mathbf{F}^\bullet$ and $\phi_f\mathbf{F}^\bullet$, respectively.

Remark 10.5. Suppose that h is an analytic function from some open neighborhood, $\mathcal{U}$ of the origin in $M = \mathbb{C}^{n+1}$. For the Lê cycles, we are interested in the case where $X = \Sigma h$ and, hence, $s = \dim_{\mathbf{0}} X = \dim_{\mathbf{0}}\Sigma h$.

Now, $\phi_h\mathbb{C}^\bullet_{\mathcal{U}}$ is positively perverse on $h^{-1}(0)$ (with no shifts), and the support of $\phi_h\mathbb{C}^\bullet_{\mathcal{U}}$ is contained in Σh. It follows that

$$\mathbf{P}^\bullet := \left(\phi_h\mathbb{C}^\bullet_{\mathcal{U}}\right)_{|\Sigma h}[n-s]$$

is positively perverse on Σh. Moreover, this complex, $\mathbf{P}^\bullet$, on Σh is such that, for all $\mathbf{x} \in \Sigma h$, the stalk cohomology, $H^i(\mathbf{P}^\bullet)_{\mathbf{x}}$ equals the reduced cohomology of the Milnor fibre of h at $\mathbf{x}$ with complex coefficients, i.e.

$$H^i(\mathbf{P}^\bullet)_{\mathbf{x}} = \widetilde{H}^i(F_{h,\mathbf{x}};\mathbb{C}).$$

In particular, the Euler characteristics satisfy

$$\chi(\mathbf{P}^\bullet)_{\mathbf{x}} = \tilde{\chi}(F_{h,\mathbf{x}}).$$

We claim now that the Morse inequalities of Theorem 3.3 are general results which apply to any positively perverse sheaf on a space. We outline this below.

In place of the Lê cycles, we have the following result for all bounded, constructible complexes $\mathbf{F}^\bullet$ on X.

Proposition/Definition 10.6. *For a generic, linear choice of coordinates $\mathbf{z} = (z_0, \dots, z_n)$ for $\mathbb{C}^{n+1}$, there exist analytic cycles, $\Lambda^i_{\mathbf{F}^\bullet,\mathbf{z}}$, in X which are purely i-dimensional, such that $\Lambda^i_{\mathbf{F}^\bullet,\mathbf{z}}$ and $V(z_0 - x_0, \dots, z_{i-1} - x_{i-1})$ intersect properly at each point $\mathbf{x} \in X$ near the origin, and such that for all $\mathbf{x}$ near the origin in X,*

$$\chi(\mathbf{F}^\bullet)_{\mathbf{x}} = \sum_{i=0}^{s} (-1)^{s-i} \left(\Lambda^i_{\mathbf{F}^\bullet,\mathbf{z}} \cdot V(z_0 - x_0, \dots, z_{i-1} - x_{i-1})\right)_{\mathbf{x}}.$$

Moreover, whenever such $\Lambda^i_{\mathbf{F}^\bullet,\mathbf{z}}$ exist, they are unique.

We call $\Lambda^i_{\mathbf{F}^\bullet,\mathbf{z}}$ *the i-th characteristic polar cycle of* $\mathbf{F}^\bullet$ *with respect to* $\mathbf{z}$. Also, we write $\lambda^i_{\mathbf{F}^\bullet,\mathbf{z}}(\mathbf{x})$ for $\left(\Lambda^i_{\mathbf{F}^\bullet,\mathbf{z}} \cdot V(z_0 - x_0, \dots, z_{i-1} - x_{i-1})\right)_{\mathbf{x}}$ and call it the *i-th characteristic polar multiplicity* (of $\mathbf{F}^\bullet$ with respect to $\mathbf{z}$).

Remark 10.7. With the above definition, and using Theorem 10.3 and Remark 10.5, we see that the Lê numbers of h are precisely the characteristic polar multiplicities of the (shifted, restricted) sheaf of vanishing cycles of h. More precisely, if we have $h : (\mathcal{U}, \mathbf{0}) \to (\mathbb{C}, 0)$, $X = \Sigma h$, $s = \dim_{\mathbf{0}} X$, and we let

$$\mathbf{P}^\bullet = \left(\phi_h \mathbb{C}^\bullet_{\mathcal{U}}\right)_{|\Sigma h} [n - s]$$

then, for generic $\mathbf{z}$, for all i, and for all $\mathbf{x} \in \Sigma h$ near the origin,

$$\Lambda^i_{\mathbf{P}^\bullet,\mathbf{z}} = \Lambda^i_{h,\mathbf{z}} \quad \text{and} \quad \lambda^i_{\mathbf{P}^\bullet,\mathbf{z}}(\mathbf{x}) = \lambda^i_{h,\mathbf{z}}(\mathbf{x}).$$

We have the following result which relates the characteristic polar multiplicities to iterated nearby and vanishing cycles with respect to generic linear forms. We concentrate our attention at the origin for simplicity.

Theorem 10.8. *Let* $\mathbf{F}^\bullet$ *be a bounded, constructible sheaf on an s-dimensional complex analytic subset,* X, *of* $\mathbb{C}^{n+1}$. *Then, for a generic linear choice of coordinates,* $\mathbf{z}$,

$$\lambda^0_{\mathbf{F}^\bullet,\mathbf{z}}(\mathbf{0}) = (-1)^{s-1}\chi(\phi_{z_0}\mathbf{F}^\bullet)_{\mathbf{0}},$$

$$\lambda^1_{\mathbf{F}^\bullet,\mathbf{z}}(\mathbf{0}) = (-1)^{s-2}\chi(\phi_{z_1}\psi_{z_0}\mathbf{F}^\bullet)_{\mathbf{0}},$$

$$\vdots$$

$$\lambda^{s-1}_{\mathbf{F}^\bullet,\mathbf{z}}(\mathbf{0}) = \chi(\phi_{z_{s-1}}\psi_{z_{s-2}}\cdots\psi_{z_0}\mathbf{F}^\bullet)_{\mathbf{0}},$$

and

$$\lambda^{s}_{\mathbf{F}^\bullet,\mathbf{z}}(\mathbf{0}) = \chi(\psi_{z_{s-1}}\psi_{z_{s-2}}\cdots\psi_{z_0}\mathbf{F}^\bullet)_{\mathbf{0}}.$$

In particular, if $\mathbf{P}^\bullet$ *is a positively perverse sheaf, then the characteristic polar cycles of* $\mathbf{P}^\bullet$ *are all non-negative, and*

$$\lambda^0_{\mathbf{P}^\bullet,\mathbf{z}}(\mathbf{0}) = \dim H^{s-1}(\phi_{z_0}\mathbf{P}^\bullet)_{\mathbf{0}},$$

$$\lambda^1_{\mathbf{P}^\bullet,\mathbf{z}}(\mathbf{0}) = \dim H^{s-2}(\phi_{z_1}\psi_{z_0}\mathbf{P}^\bullet)_{\mathbf{0}},$$

$$\vdots$$

$$\lambda^{s-1}_{\mathbf{P}^\bullet,\mathbf{z}}(\mathbf{0}) = \dim H^{0}(\phi_{z_{s-1}}\psi_{z_{s-2}}\cdots\psi_{z_0}\mathbf{P}^\bullet)_{\mathbf{0}},$$

and

$$\lambda^{s}_{\mathbf{P}^\bullet,\mathbf{z}}(\mathbf{0}) = \dim H^{0}(\psi_{z_{s-1}}\psi_{z_{s-2}}\cdots\psi_{z_0}\mathbf{P}^\bullet)_{\mathbf{0}}.$$

We will now describe, for a positively perverse sheaf, $\mathbf{P}^\bullet$, how to obtain a chain complex with cohomology equal to that of the stalk cohomology of $\mathbf{P}^\bullet$ at $\mathbf{0}$ and in which the dimensions of the terms of the complex are given by the characteristic polar multiplicities.

If $\mathbf{Q}^\bullet$ is a positively perverse sheaf on an m-dimensional germ, $(Y, \mathbf{0})$, which is embedded in some affine space then, for a generic linear form, z_i, the sheaf of vanishing cycles, $\phi_{z_i}\mathbf{Q}^\bullet$, is a positively perverse sheaf on $Y \cap V(z_i)$ which has the origin as an isolated point in its support. Therefore, the stalk cohomology of $\phi_{z_i}\mathbf{Q}^\bullet$ at the origin is zero in all dimensions except possibly in dimension $m-1$.

From the long exact sequence on stalk cohomology which is associated to the distinguished triangle defining $\phi_{z_i}\mathbf{Q}^\bullet$, we find that, for $k \leqslant m-2$, $H^k(\mathbf{Q}^\bullet)_{\mathbf{0}} \cong H^k(\psi_{z_i}\mathbf{Q}^\bullet)_{\mathbf{0}}$ and we have the exact sequence

$$0 \to H^{m-1}(\mathbf{Q}^\bullet)_{\mathbf{0}} \to H^{m-1}(\psi_{z_i}\mathbf{Q}^\bullet)_{\mathbf{0}} \xrightarrow{\beta}$$

$$H^{m-1}(\phi_{z_i}\mathbf{Q}^\bullet)_{\mathbf{0}} \xrightarrow{\gamma} H^{m}(\mathbf{Q}^\bullet)_{\mathbf{0}} \to 0.$$

Applying this repeatedly with $\mathbf{Q}^\bullet$ replaced by each of the

$$\phi_{z_{k-1}}\psi_{z_{k-2}}\cdots\psi_{z_0}\mathbf{P}^\bullet,$$

we obtain the exact sequences

$$0 \to H^{s-1}(\mathbf{P}^\bullet)_\mathbf{0} \to H^{s-1}(\psi_{z_0}\mathbf{P}^\bullet)_\mathbf{0} \xrightarrow{\beta_{s-1}} H^{s-1}(\phi_{z_0}\mathbf{P}^\bullet)_\mathbf{0} \xrightarrow{\gamma_{s-1}} H^s(\mathbf{P}^\bullet)_\mathbf{0} \to 0$$

$$0 \to H^{s-2}(\psi_{z_0}\mathbf{P}^\bullet)_\mathbf{0} \to H^{s-2}(\psi_{z_1}\psi_{z_0}\mathbf{P}^\bullet)_\mathbf{0} \xrightarrow{\beta_{s-2}} H^{s-2}(\phi_{z_1}\psi_{z_0}\mathbf{P}^\bullet)_\mathbf{0} \xrightarrow{\gamma_{s-2}} H^{s-1}(\psi_{z_0}\mathbf{P}^\bullet)_\mathbf{0} \to 0$$

$$\vdots$$

$$0 \to H^1(\psi_{z_{s-3}}\psi_{z_{s-2}}\cdots\psi_{z_0}\mathbf{P}^\bullet)_\mathbf{0} \to H^1(\psi_{z_{s-2}}\psi_{z_{s-3}}\cdots\psi_{z_0}\mathbf{P}^\bullet)_\mathbf{0} \xrightarrow{\beta_1} H^1(\phi_{z_{s-2}}\psi_{z_{s-3}}\cdots\psi_{z_0}\mathbf{P}^\bullet)_\mathbf{0} \xrightarrow{\gamma_1} H^2(\psi_{z_{s-3}}\psi_{z_{s-2}}\cdots\psi_{z_0}\mathbf{P}^\bullet)_\mathbf{0} \to 0$$

$$0 \to H^0(\psi_{z_{s-2}}\psi_{z_{s-3}}\cdots\psi_{z_0}\mathbf{P}^\bullet)_\mathbf{0} \to H^0(\psi_{z_{s-1}}\psi_{z_{s-2}}\cdots\psi_{z_0}\mathbf{P}^\bullet)_\mathbf{0} \xrightarrow{\beta_0} H^0(\phi_{z_{s-1}}\psi_{z_{s-2}}\cdots\psi_{z_0}\mathbf{P}^\bullet)_\mathbf{0} \xrightarrow{\gamma_0} H^1(\psi_{z_{s-2}}\psi_{z_{s-3}}\cdots\psi_{z_0}\mathbf{P}^\bullet)_\mathbf{0} \to 0$$

With the above notation, we now have

Theorem 10.9.

$$0 \to H^0(\psi_{z_{s-1}}\psi_{z_{s-2}}\cdots\psi_{z_0}\mathbf{P}^\bullet)_\mathbf{0} \xrightarrow{\beta_0} H^0(\phi_{z_{s-1}}\psi_{z_{s-2}}\cdots\psi_{z_0}\mathbf{P}^\bullet)_\mathbf{0} \xrightarrow{\beta_1\circ\gamma_0} H^1(\phi_{z_{s-2}}\psi_{z_{s-3}}\cdots\psi_{z_0}\mathbf{P}^\bullet)_\mathbf{0} \xrightarrow{\beta_2\circ\gamma_1} \cdots \xrightarrow{\beta_{s-2}\circ\gamma_{s-3}} H^{s-2}(\phi_{z_1}\psi_{z_0}\mathbf{P}^\bullet)_\mathbf{0} \xrightarrow{\beta_{s-1}\circ\gamma_{s-2}} H^{s-1}(\phi_{z_0}\mathbf{P}^\bullet)_\mathbf{0} \to 0$$

is a chain complex with cohomology equal to the stalk cohomology of $\mathbf{P}^\bullet$ *at the origin.*

Proof. Given the statement, the proof of this is completely trivial – one uses repeatedly that the γ_i are surjections. □

In the case where X is the s-dimensional critical locus of a function $h : (\mathcal{U}, \mathbf{0}) \to (\mathbb{C}, 0)$ and $\mathbf{P}^\bullet = (\phi_h \mathbb{C}^\bullet_{\mathcal{U}})_{|\Sigma h}[n-s]$, Remark 10.7 allows us to conclude from Theorem 10.9 that we have

Corollary 10.10. *There is a chain complex*

$$0 \to \mathbb{C}^{\lambda^0_h} \to \mathbb{C}^{\lambda^1_h} \cdots \to \mathbb{C}^{\lambda^{n-1}_h} \to \mathbb{C}^{\lambda^n_h} \to 0$$

with cohomology equal to the reduced cohomology of the Milnor fibre of h at the origin, where we have written λ^i_h for $\lambda^i_{h,\mathbf{z}}(\mathbf{0})$, and the $\mathbb{C}^{\lambda^i_h}$ term stands in degree $n-i$.

This gives a very general explanation for the Morse inequalities which appear in Theorem 3.3.

Remark 10.11. The terms in the chain complex of 10.9 are sums of Morse groups of strata "counted with multiplicity" – where the "multiplicity" comes from the polar multiplicities of the strata. As we shall see shortly, if $\{S_\alpha\}$ is a Whitney stratification with respect to which $\mathbf{P}^\bullet$ is constructible and if we let A_α denote the Morse group of S_α with respect to $\mathbf{P}^\bullet$, then

$$H^{s-k}(\phi_{z_{k-1}} \psi_{z_{k-2}} \cdots \psi_{z_0} \mathbf{P}^\bullet)_0 = \sum_\alpha A_\alpha \otimes \mathbb{C}^{p_{\alpha,k-1}},$$

where $p_{\alpha,k-1}$ denotes the multiplicity of the $(k-1)$-dimensional polar variety of $\overline{S_\alpha}$ at the origin.

In the very special case where $\overline{S_\alpha}$ is smooth at the origin for every stratum, the polar multiplicities vanish – except for the multiplicity of the polar variety whose dimension matchs that of S_α; this multiplicity is obviously equal to 1. Hence, in this case, the terms of the chain complex in 10.9 are

$$H^{s-k}(\phi_{z_{k-1}} \psi_{z_{k-2}} \cdots \psi_{z_0} \mathbf{P}^\bullet)_0 = \sum_{dim S_\alpha = k-1} A_\alpha$$

and

$$H^0(\psi_{z_{s-1}} \psi_{z_{s-2}} \cdots \psi_{z_0} \mathbf{P}^\bullet)_0 = \sum_{dim S_\alpha = s} A_\alpha.$$

We now wish to describe a related result - one which shows the connection between the characteristic polar cycles of a complex of sheaves and the characteristic cycle of the complex (hence, the first part of the term "characteristic polar cycles"). This enables us to give a description of the characteristic polar cycles which involves the absolute polar varieties of Lê and Teissier (hence, the rest of the term). We are also able then to give another very algebraic description of the Lê cycles of a hypersurface singularity.

We will first quickly review the definition and some of the results on the characteristic cycle of a constructible sheaf. More complete references are [**BDK**], [**Gi**], [**K-S1**], [**K-S2**], [**L-M**], [**Mac2**], and [**Sab**]; note, however, that many of these references are in the language of holonomic $\mathcal{D}$-modules and not the language of perverse sheaves.

We continue with $\mathbf{F}^\bullet$ being a bounded, constructible sheaf on an s-dimensional complex analytic space, X, embedded in some affine space $M := \mathbb{C}^{n+1}$. Later, we want to projectivize the conormal bundle, $T^*_X M$, so we require that the codimension of X in M be greater than or equal to 1, i.e. $s \leqslant n$.

Let $\{S_\alpha\}$ be any Whitney stratification of X (with connected strata) with respect to which $\mathbf{F}^\bullet$ is constructible. Then, the *characteristic cycle*, $Ch(\mathbf{F}^\bullet)$, *of* $\mathbf{F}^\bullet$ in T^*M is the linear combination $\sum_\alpha m_\alpha \overline{T^*_{S_\alpha} M}$, where $\overline{T^*_{S_\alpha} M}$ denotes the closure of the conormal bundle to S_α in T^*M and the m_α are integers given by

$$(10.12) \qquad m_\alpha := (-1)^{s-1}\chi(\phi_{f_{|X}}\mathbf{F}^\bullet)_{\mathbf{x}} = (-1)^{s-d-1}\chi(\phi_{f_{|N}}\mathbf{F}^\bullet{}_{|N})_{\mathbf{x}}$$

for all points $\mathbf{x}$ in the d-dimensional stratum S_α, with normal slice N to S_α at $\mathbf{x}$, and any $f : M, x \to \mathbb{C}, 0$ such that $d_{\mathbf{x}}f$ is a non-degenerate covector at $\mathbf{x}$ (with respect to our fixed stratification; see [**G-M2**]) and $f_{|S_\alpha}$ has a Morse singularity at $\mathbf{x}$. This cycle is independent of all the choices made.

The fact that $d_{\mathbf{x}}f$ is non-degenerate is equivalent to saying that if S_α is the stratum containing $\mathbf{x}$, then $(\mathbf{x}, d_{\mathbf{x}}f)$ is not in $\overline{T^*_{S_\beta} M}$ for $S_\alpha \neq S_\beta$; thus, $(\mathbf{x}, d_{\mathbf{x}}f)$ is a smooth point of $Ch(\mathbf{F}^\bullet)$.

Hence, for such an $\mathbf{x}$ and f,

$$(Ch(\mathbf{F}^\bullet) \cdot Im\ df)_{(\mathbf{x}, d_{\mathbf{x}}f)} = \left(m_\beta \overline{T^*_{S_\beta} M} \cdot Im\ df\right)_{(\mathbf{x}, d_{\mathbf{x}}f)} = m_\beta\ ,$$

since

$$\left(\overline{T^*_{S_\beta} M} \cdot Im\ df\right)_{(\mathbf{x}, d_{\mathbf{x}}f)} = \left(T^*_{S_\beta} M \cdot Im\ df\right)_{(\mathbf{x}, d_{\mathbf{x}}f)} = 1$$

as $f_{|S_\beta}$ is Morse.

Therefore, at a smooth point $(\mathbf{x}, \xi) \in Ch(\mathbf{F}^\bullet)$, if f is Morse with respect to $\{S_\alpha\}$, with $f(\mathbf{x}) = 0$ and $d_{\mathbf{x}}f = \xi$, then

$$(10.13) \qquad (-1)^{s-1}\chi(\phi_f \mathbf{F}^\bullet)_{\mathbf{x}} = (Ch(\mathbf{F}^\bullet) \cdot Im\ df)_{(\mathbf{x}, \xi)}\,.$$

Consider now the case of a perverse sheaf, $\mathbf{P}^\bullet$. Then, $Ch(\mathbf{P}^\bullet)$ is a non-negative cycle, where the m_α are given by

$$m_\alpha = \dim H^{s-1}(\phi_{f_{|X}}\mathbf{P}^\bullet)_{\mathbf{x}}.$$

In this case, the result of Ginsburg [**Gi**], Sabbah [**Sab**], and Lê [**Lê2**] is that 10.13 holds at any point $(\mathbf{x}, \xi) \in Ch(\mathbf{P}^\bullet)$ and for all f such that $f(\mathbf{x}) = 0, d_{\mathbf{x}} f = \xi$, and $(\mathbf{x}, \xi)$ is an isolated point in the intersection $Ch(\mathbf{P}^\bullet) \cap Im\ df$.

Let $\mathbb{P}(T^*M) \cong \mathbb{C}^{n+1} \times \mathbb{P}^n$ denote the projectivized cotangent bundle. Let η be the projection from $\mathbb{P}(T^*M)$ to $M = \mathbb{C}^{n+1}$.

Since $Ch(\mathbf{F}^\bullet)$ is a conical cycle in T^*M, we may consider its projectivization, $\mathbb{P}(Ch(\mathbf{F}^\bullet))$, in $\mathbb{P}(T^*M)$.

For a given choice of coordinates, $(z_0, \dots, z_n)$, on $\mathbb{C}^{n+1}$, let L^i be the i-dimensional subspace $V(z_0, \dots, z_{n-i})$; for each $\mathbf{x} \in M$, this determines a corresponding subspace of $T_{\mathbf{x}}M$ – we denote this subspace by L^i also. Let $\check{L}^i$ denote the subbundle of T^*M defined by

$$\check{L}^i = \left\{(\mathbf{x}, \xi) \in T^*M \mid \xi(L^i) = 0\right\} =$$

$$\left\{(\mathbf{x}, a_0 d_{\mathbf{x}} z_0 + \cdots + a_{n-i} d_{\mathbf{x}} z_{n-i}) \mid (a_0, \dots, a_{n-i}) \in \mathbb{C}^{n-i+1}\right\} \cong M \times \mathbb{C}^{n-i+1}.$$

We wish to consider the projectivization $\mathbb{P}(\check{L}^i) \cong M \times \mathbb{P}^{n-i}$ in $\mathbb{P}(T^*M)$.

With these notations, we have:

Theorem 10.14. *For a generic linear choice of coordinates, $\mathbf{z}$, the $(n-i)$-dimensional characteristic polar cycle, $\Lambda^{n-i}_{\mathbf{F}^\bullet, \mathbf{z}}$, of a constructible sheaf $\mathbf{F}^\bullet$ is equal to*

$$\eta_*\left(\mathbb{P}(Ch(\mathbf{F}^\bullet)) \cdot \mathbb{P}(\check{L}^i)\right),$$

where η_* denotes the push-forward of cycles (not cycle classes).

Now, $Ch(\mathbf{F}^\bullet) = \sum_\alpha m_\alpha \overline{T^*_{S_\alpha} M}$ for all Whitney stratifications $\{S_\alpha\}$ for X with respect to which $\mathbf{F}^\bullet$ is constructible. By considering each component of the characteristic cycle separately and comparing the result of the above theorem with the definition of the (absolute) polar varieties of Lê and Teissier [**L-T2**], we obtain:

Corollary 10.15. *Using the notation above, for all k,*

$$\Lambda^k_{\mathbf{F}^\bullet, \mathbf{z}} = \sum_\alpha m_\alpha \Gamma^k_{\mathbf{z}}(\overline{S_\alpha}),$$

where $\Gamma^k_{\mathbf{z}}$ denotes the polar cycle of dimension k (not codimension) with respect to the flag given by the L^i above.

Note that, in this corollary, $\Gamma^0_{\mathbf{z}}(\overline{S_\alpha})$ will normally be the zero-cycle unless S_α is a zero-dimensional stratum.

Remark 10.16. We wish to clarify exactly what Corollary 10.15 says.

If $\dim \overline{S_\alpha} = k$, then $\Gamma^k_{\mathbf{z}}(\overline{S_\alpha})$ simply equals $\overline{S_\alpha}$ with its reduced structure. Hence, the k-dimensional characteristic polar cycle has a "fixed part" which is independent of the choice of coordinates $\mathbf{z}$; this fixed part consists of the sum of $m_\alpha\left[\overline{S_\alpha}\right]$, where only those strata with non-zero m_α's contribute to the sum.

However, there is a "non-fixed part" to $\Lambda^k_{\mathbf{F}^\bullet, \mathbf{z}}$; as a set, this moving part consists of the k-dimensional absolute polar varieties of the fixed parts of higher-dimensional characteristic polar cycles.

In particular, this is the case for the Lê cycles. We saw an example of this phenomenon in the FM cone singularity (Example 2.4).

We will now consider characteristic polar multiplicities not with respect to one, fixed coordinate system but with respect to coordinate systems which are allowed to vary with the point under consideration, so that we use coordinates which are generic at each point. From the results of [**Te4**], for a generic choice of $\mathbf{z}$ at $\mathbf{x}$, the intersection number

$$\left(\Gamma^k_{\mathbf{z}}(\overline{S_\alpha}) \cdot V(z_0 - x_0, \dots, z_{k-1} - x_{k-1})\right)_{\mathbf{x}}$$

is equal to the k-dimensional polar multiplicity of $\overline{S_\alpha}$ at $\mathbf{x}$; this is an analytic invariant of the germ of the space $\overline{S_\alpha}$ at $\mathbf{x}$. Moreover, the polar multiplicities of a space are constant along Whitney strata (again, see [**Te4**]).

We will let $\Lambda^i_{\mathbf{F}^\bullet}$ denote the i-th characteristic polar cycle of $\mathbf{F}^\bullet$ as a germ at $\mathbf{x}$ with respect to coordinates which are generic at the point $\mathbf{x}$ and we denote the corresponding characteristic polar multiplicity by $\lambda^i_{\mathbf{F}^\bullet}(\mathbf{x})$.

We now wish to investigate the consequences of Teissier's result on the constancy of the polar multiplicities implying Whitney conditions [**Te4**].

Recall that for a positively perverse sheaf, $\mathbf{P}^\bullet$, the m_α which appear in the characteristic cycle are all non-negative. This allows us to easily go from Teissier's result to:

Theorem 10.17. *Let $\mathbf{P}^\bullet$ be a positively perverse sheaf on an s-dimensional complex analytic space, X. Assume that X is embedded in some affine space $M := \mathbb{C}^{n+1}$. Let $\{S_\alpha\}$ be any Whitney stratification of X (with connected strata) with respect to which $\mathbf{P}^\bullet$ is constructible, and let $Ch(\mathbf{P}^\bullet) = \sum_\alpha m_\alpha \overline{T^*_{S_\alpha} M}$. Finally, let N be a submanifold of X.*

If the generic characteristic polar multiplicities of $\mathbf{P}^\bullet$ are all constant along N, then the pair (S_α, N) satisfies Whitney conditions for every α for which $m_\alpha \neq 0$.

We shall conclude by considering the ramifications of the above general results in the Lê cycle case, i.e. the case where X is the s-dimensional critical locus of a function $h : (\mathcal{U}, \mathbf{0}) \to (\mathbb{C}, 0)$ and $\mathbf{P}^\bullet = \left(\phi_h \mathbb{C}^\bullet_{\mathcal{U}}\right)_{|_{\Sigma h}} [n-s]$.

The result of Lê and Mebkhout in [**L-M**] is that the cycle $\mathbb{P}(Ch(\mathbf{P}^\bullet))$ is equal to the exceptional divisor, E, of the blow-up of the Jacobian ideal of h in $\mathcal{U}$. Combining this with Theorem 10.14, and using the notation from that theorem, we obtain a result of T. Gaffney:

Theorem 10.18. *For a generic choice of coordinates,* $\mathbf{z}$, *the* $(n-i)$*-dimensional Lê cycle,* $\Lambda^{n-i}_{h,\mathbf{z}}$, *of* h *is equal to* $\eta_* \left(E \cdot \mathbb{P}(\check{L}^i)\right)$.

Gaffney's proof of the above statement is very algebraic and uses Segre classes; a similar, more general, proof also seems to be contained in [**Gas1**] and [**Gas2**]. It is also possible to look at the defining equations of E on affine patches in $\mathbb{P}(T^*M)$ and prove directly that $\eta_* \left(E \cdot \mathbb{P}(\check{L}^i)\right)$ coincides with the definition of the Lê cycles.

In keeping with our earlier notation, when we omit the subscript referring to the coordinates, we mean that one should use coordinates which are generic at the point under consideration. In particular, $\lambda^i_h(\mathbf{p})$ denotes the i-th Lê number of h at $\mathbf{p}$ with respect to coordinates that are generic at $\mathbf{p}$ (which, by the above, equals the i-th characteristic polar multiplicity of the sheaf of vanishing cycles of h).

Teissier's results on the polar multiplicities [**Te4**] allow us to conclude:

Theorem 10.19. *The generic Lê numbers,* λ^i_h, *of an analytic function* h *are analytic invariants. Moreover, these generic Lê numbers are constant along the strata of any Whitney stratification of* $V(h)$.

Theorem 9.4 and Remark 9.5 immediately yield:

Theorem 10.20. *Let* N *be a* 1*-dimensional submanifold of* $\mathcal{U}$ *along which all the generic Lê numbers,* λ^i_h, *are constant, then the stalk cohomology of the sheaf of vanishing cycles* $\phi_h \mathbb{C}^\bullet_M$ *is constant along* N.

However, given the statement of Theorem 10.17, we might hope that keeping the generic Lê numbers constant along N would actually imply that the critical locus of h has a Whitney stratification with N as a stratum and that, along N, the cohomology of $\phi_h\mathbb{C}^\bullet_M$ is locally constant. It is important to note that we are not suggesting that the smooth part of the hypersurface $V(h)$ can be included in this Whitney stratification; for instance, if h has a smooth 1-dimensional critical locus, then the hoped-for result is automatic.

In fact, for a one-dimensional singularity, one does not even have to explicitly require λ^1_h to be constant to obtain:

Proposition 10.21. *If the critical locus, Σh, of h is one-dimensional at the origin and $\lambda^0_h(\mathbf{0}) = 0$, then the critical locus at the origin has a single, smooth component along which the sheaf of vanishing cycles is locally constant.*

Proof. We have only to show that the critical locus has a single smooth component through the origin.

As $\lambda^0_h(\mathbf{0}) = 0$, it follows from the definition of the Lê numbers that h has no generic relative polar curve at the origin. It is immediate that the Milnor number at the origin of a generic hyperplane slice of h equals $\lambda^1_h(\mathbf{0})$. This is precisely the situation of [**Lê5**] – the conclusion follows. □

By combining 10.21 with 10.17, it is not difficult to show that

Theorem 10.22. *If* $\dim_{\mathbf{0}}\Sigma h = 2$ *and N is a smooth curve through the origin which is contained in Σh and along which the generic Lê numbers are constant, then – near the origin – N may be chosen as a Whitney stratum in the space Σh and the sheaf of vanishing cycles, $\phi_h\mathbb{C}^\bullet_M$, is locally constant along N.*

The barrier to proving such a result for higher-dimensional critical loci is that we know of no way to know when the various strata occur with non-zero multiplicity in the characteristic cycle of the sheaf of vanishing cycles; such knowledge is crucial in order to apply Theorem 10.17.

APPENDIX:

PRIVILEGED NEIGHBORHOODS AND LIFTING MILNOR FIBRATIONS

In this appendix, we prove a number of very technical results. These results tell us when we can use certain types of "nice" neighborhoods to define the Milnor fibre (at least, up to homotopy), and give conditions under which Milnor fibrations remain constant in a parameterized family. Lê numbers and cycles do not appear here, though we will use the relative polar curve.

Throughout, for convenience, we concentrate our attention at the origin. Let $\mathcal{U}$ be an open neighborhood of the origin in some $\mathbb{C}^{n+1}$ and let $h : (\mathcal{U}, \mathbf{0}) \to (\mathbb{C}, 0)$ be an analytic function.

In what sense the Milnor fibre and Milnor fibration of h are well-defined has been discussed in a number of places (see, for instance, [**Se-Th**]). If one is primarily interested in the ambient, local topology of the hypersurface $V(h)$ defined by h, then "the" Milnor fibre is only well-defined up to homotopy-type [**Lê6**]. Thus, we may make the weakest possible definition of the Milnor fibre of h at the origin as a homotopy-type:

Definition/Proposition A.1. A *system of Milnor neighborhoods* for h at the origin is a fundamental system of neighborhoods, $\{C_\alpha\}$, at the origin in $\mathcal{U}$ such that for all $C_\alpha \subseteq C_\beta$, there exists $\epsilon > 0$ such that for all complex ξ with $0 < |\xi| < \epsilon$, we have that the inclusion $C_\alpha \cap V(h-\xi) \hookrightarrow C_\beta \cup V(h-\xi)$ is a homotopy-equivalence. The *standard system of Milnor neighborhoods* for h at the origin is just the set of closed balls of sufficiently small radius centered at the origin (this system is independent of h except for how small the radii must be).

If $\{C_\alpha\}$ is a system of Milnor neighborhoods for h at the origin, then for each C_α there exists $\epsilon > 0$ such that the homotopy-type of $C_\alpha \cap V(h-\xi)$ is independent of the complex number ξ chosen as long as $0 < \xi < \epsilon$. Moreover, this homotopy-type is independent of the choice of the particular C_α and is, in fact, independent of the choice of the system of Milnor neighborhoods.

Proof. The proof is standard. Let $\{C_\alpha\}$ be a system of Milnor neighborhoods for h at the origin. We shall compare it with the standard system. Select any C_β. Now, pick C_α, B_η, and B_δ such that B_η and B_δ are in the standard system of Milnor neighborhoods for h at the origin and such that $B_\delta \subseteq C_\alpha \subseteq B_\eta \subseteq C_\beta$. We may certainly pick $\epsilon > 0$ such that, for all complex ξ with $0 < |\xi| < \epsilon$, the inclusion $C_\alpha \cap V(h-\xi) \hookrightarrow C_\beta \cap V(h-\xi)$ and the inclusion $B_\delta \cap V(h-\xi) \hookrightarrow B_\eta \cap V(h-\xi)$ are both homotopy-equivalences. It follows that the inclusion $C_\alpha \cap V(h-\xi) \hookrightarrow B_\eta \cap V(h-\xi)$ is a homotopy-equivalence for all small $\xi \neq 0$ and thus, as the homotopy-type of $B_\eta \cap V(h-\xi)$ is independent of ξ, so is that of $C_\alpha \cap V(h-\xi)$. The conclusion follows immediately. $\square$

It is sometimes more convenient to prove that $C_\alpha \cap V(h-\xi) \hookrightarrow C_\beta \cap V(h-\xi)$ is a homotopy-equivalence whenever C_α is contained in the *interior* of C_β. It is easy to see by the proof above that this is enough to show that the system is a system of Milnor neighborhoods.

A system of Milnor neighborhoods allows one to discuss the Milnor fibre up to homotopy. However, one frequently wishes to use stratified, differential techniques to study the Milnor fibre and, hence, one would like for the Milnor fibre to have the structure of a smooth, compact manifold with (stratified) boundary and would also like to have some control over what happens on the boundary as one moves through a family of singularities. Furthermore, one would like to have a notion of the Milnor *fibration* – at least up to fibre-homotopy-type.

To gain this additional structure, we will use two types of (complex analytic) stratifications. One is the well-known Whitney stratification [**G-M2**], [**Mat**], [**Th**]. The second is a *good stratification*, as defined in 1.24. Note that any refinement of a good stratification which does not refine the smooth stratum is automatically a good stratification. This fact will be very useful when combined with the following proposition, which is Theorem 18.11 of [**W**] (or just a small portion of Theorem 1.7 of [**G-M2**]).

Proposition A.2. *Let X be an analytic subset of $\mathbb{C}^N$ and let Y be an analytic subset of X. Suppose that $\mathcal{D}$ and $\mathcal{F}$ are analytic stratifications for X and Y, respectively. Then, there exists an analytic stratification, $\mathcal{L}$, of X which is a common refinement of both $\mathcal{D}$ and $\mathcal{F}$, i.e. every stratum of $\mathcal{L}$ is contained in a stratum of $\mathcal{D}$, Y is a union of strata of $\mathcal{L}$, and every stratum of $\mathcal{L}$ which is contained in Y is contained in a stratum of $\mathcal{F}$.*

The $\mathcal{L}$ above is sometimes referred to as a refinement of $\mathcal{D}$ adapted to Y.

(The reader should note that when Goresky and MacPherson use the term "stratification", they mean that the Whitney conditions are satisfied. Hence, their Theorem 1.7 actually allows us to pick a common analytic, Whitney refinement.)

We now generalize the notion of a privileged polydisc as given in [**Lê3**]. This definition should be compared with [**L-T1**, 2.2.3].

Definition A.3. Let $\mathfrak{G}$ be a good stratification for h at the origin. A fundamental system of neighborhoods, $\{C_\alpha\}$, at the origin in $\mathcal{U}$ is a *system of privileged neighborhoods for h at $\mathbf{0}$ with respect to $\mathfrak{G}$* if and only if

i) $\{C_\alpha\}$ is a system of compact, Milnor neighborhoods for h at the origin;

and, for each C_α, there is an associated Whitney stratification, $\mathcal{S}_\alpha$, of C_α such that

ii) the interior of C_α, $\overset{\circ}{C}_\alpha$, in $\mathcal{U}$ is a stratum in $\mathcal{S}_\alpha$;

iii) C_α equals the closure of $\overset{\circ}{C}_\alpha$ in $\mathcal{U}$;

By ii) and iii) and the condition of the frontier, the boundary of C_α, ∂C_α, is a union of Whitney strata, and we make the final requirement:

iv) the boundary strata of each C_α transversely intersect all the strata of $\mathfrak{G}$.

A fundamental system of neighborhoods, $\mathcal{C} = \{C_\alpha\}$, is a *system of privileged neighborhoods for h at* $\mathbf{0}$ if and only if there exists a good stratification, $\mathfrak{G}$, for h at $\mathbf{0}$ such that $\mathcal{C}$ is a system of privileged neighborhoods for h at $\mathbf{0}$ with respect to $\mathfrak{G}$.

A fundamental system of neighborhoods, $\mathcal{C} = \{C_\alpha\}$, satisfying i), ii), and iii) above is a *system of weakly privileged neighborhoods for h at* $\mathbf{0}$ if and only if for each C_α, for all small $\xi \neq 0$, $V(h - \xi)$ transversely intersects the boundary strata of C_α. We shall see below that a system of privileged neighborhoods is automatically a system of weakly privileged neighborhoods.

A fundamental system of neighborhoods, $\mathcal{C} = \{C_\alpha\}$, is a *universal system of privileged neighborhoods for h at* $\mathbf{0}$ if and only if for every good stratification, $\mathfrak{G}$, for h at $\mathbf{0}$, there exists an open neighborhood W of the origin such that $\{C_\alpha \in \mathcal{C} \mid C_\alpha \subseteq W\}$ is a system of privileged neighborhoods for h at $\mathbf{0}$ with respect to $\mathfrak{G}$.

One should note that the set of closed balls centered at the origin is a universal system of privileged neighborhoods for h, regardless of the function h – this is a very "universal" system, and this may seem like the more natural notion. This, however, seems to be too restrictive. Universal for h simply means that, locally, the fundamental system is privileged independent of the choice of good stratification for the particular function h.

Proposition A.4. *Suppose that $\mathcal{C} = \{C_\alpha\}$ is a system of privileged neighborhoods for h at $\mathbf{0}$. Then, $\mathcal{C} = \{C_\alpha\}$ is a system of weakly privileged neighborhoods for h at $\mathbf{0}$ and, hence, for all C_α, for all small $\delta > 0$, $C_\alpha \cap h^{-1}(\partial\mathbb{D}_\delta) \xrightarrow{h} \partial\mathbb{D}_\delta$ is a proper, stratified submersion and is thus a locally trivial fibration. The fibre-homotopy-type of this fibration is independent of the choice of the system of weakly privileged neighborhoods, $\mathcal{C}$, for h, the choice of C_α, and the choice of small $\delta > 0$.*

Proof. The proof is essentially that of Lê in [**Lê 4**]. Let $\mathfrak{G}$ be a good stratification for h at the origin with respect to which $\mathcal{C}$ is a system of privileged neighborhoods. Pick a C_α in $\mathcal{C}$. We shall actually show that there exists $\epsilon > 0$ such that

$C_\alpha \cap h^{-1}(\overset{\circ}{\mathbb{D}}_\epsilon - 0) \xrightarrow{h} \overset{\circ}{\mathbb{D}}_\epsilon - 0$ is a proper, stratified submersion. It follows that, for all δ with $0 < \delta < \epsilon$, $C_\alpha \cap h^{-1}(\partial\mathbb{D}_\delta) \xrightarrow{h} \partial\mathbb{D}_\delta$ is a proper, stratified submersion and, hence, a locally trivial fibration with fibre-homotopy-type independent of the choice of δ. This certainly shows that $\mathcal{C}$ is a system of weakly privileged neighborhoods for h at $\mathbf{0}$.

Suppose to the contrary that no matter how small we choose $\epsilon > 0$ it is not the case that $C_\alpha \cap h^{-1}(\overset{\circ}{\mathbb{D}}_\epsilon - 0) \xrightarrow{h} \overset{\circ}{\mathbb{D}}_\epsilon - 0$ is a proper, stratified submersion. As each C_α is compact, clearly this map is always proper. So, by the local finiteness of the stratification, there must exist a single Whitney stratum, S, of C_α and a sequence of points $\mathbf{p}_i \in S$ such that the $\mathbf{p}_i$ converge to some point $\mathbf{p} \in V(h)$, $T_{\mathbf{p}_i}S$ converges to some T, $T_{\mathbf{p}_i}V(h - h(\mathbf{p}_i))$ converges to some $\mathcal{T}$, and $T_{\mathbf{p}_i}S \subseteq T_{\mathbf{p}_i}V(h - h(\mathbf{p}_i))$. Let G denote the good stratum of $\mathfrak{G}$ which contains $\mathbf{p}$ and let R denote the Whitney stratum of C_α which contains $\mathbf{p}$.

As $T_{\mathbf{p}_i}S \subseteq T_{\mathbf{p}_i}V(h - h(\mathbf{p}_i))$, we must have that $T \subseteq \mathcal{T}$. By the Thom condition, $T_{\mathbf{p}}G \subseteq \mathcal{T}$. By Whitney's condition a), $T_{\mathbf{p}}R \subseteq T$. Hence, $T_{\mathbf{p}}R$ and $T_{\mathbf{p}}G$ are both contained in $\mathcal{T}$ – a contradiction as R and G intersect transversely.

Thus, there exists $\epsilon > 0$ such that for all δ with $0 < \delta < \epsilon$, $C_\alpha \cap h^{-1}(\partial\mathbb{D}_\delta) \xrightarrow{h} \partial\mathbb{D}_\delta$ is a proper, stratified submersion and, hence, a locally trivial fibration with fibre-homotopy-type independent of the choice of δ. To see that the fibre-homotopy-type is independent of the choice of $\mathcal{C}$ and the choice of C_α, one may once again compare with the standard system of Milnor neighborhoods and then use the theorem of Dold [**Hu**, p.209], since we know that the inclusion of each fibre is a homotopy-equivalence by the proof of A.1. We leave the details to the reader. □

Note that we have the implications: $\{C_\alpha\}$ is a universal system $\Rightarrow$ for all good stratifications $\mathfrak{G}$, $\{C_\alpha\}$ is a privileged system with respect to $\mathfrak{G} \Rightarrow \{C_\alpha\}$ is a privileged system $\Rightarrow \{C_\alpha\}$ is a weakly privileged system $\Rightarrow \{C_\alpha\}$ is a Milnor system.

Definition A.5. If $\mathcal{C} = \{C_\alpha\}$ is a system of Milnor neighborhoods for h at $\mathbf{0}$, then a *Milnor pair* for h at $\mathbf{0}$ is a pair $(C_\alpha, \overset{\circ}{\mathbb{D}}_\delta)$ such that for all $\xi \in \overset{\circ}{\mathbb{D}}_\delta - 0$, $C_\alpha \cap V(h - \xi)$ has the homotopy-type of the Milnor fibre. If, in addition, $\mathcal{C}$ is a system of weakly privileged neighborhoods, then we also make the requirement that $C_\alpha \cap h^{-1}(\partial\mathbb{D}_\delta) \xrightarrow{h} \partial\mathbb{D}_\delta$ is a proper, stratified submersion.

We now wish to consider an analytic function $f : (\overset{\circ}{\mathbb{D}} \times \mathcal{U}, \overset{\circ}{\mathbb{D}} \times \mathbf{0}) \to (\mathbb{C}, 0)$ where $\overset{\circ}{\mathbb{D}}$ is an open complex disc centered at the origin and $\mathcal{U} \subseteq \mathbb{C}^{n+1}$. We use the coordinates $(t, z_0, \ldots, z_n)$ for $\overset{\circ}{\mathbb{D}} \times \mathcal{U}$. We distinguish the t-coordinate because we will either be considering the particular hyperplane slice $V(t)$ or because we will be interested in the family $f_t(z_0, \dots, z_n) := f(t, z_0, \dots, z_n)$.

Proposition A.6. *Suppose that $V(t)$ is prepolar for f at the origin with respect to a good stratification $\mathfrak{G}$, and let $\{C_\alpha\}$ be a system of privileged neighborhoods with respect to the good stratification $\mathfrak{G} \cap V(t)$ for $f_{|_{V(t)}}$. Then, there exits an open neighborhood, W, of the origin in $V(t)$ such that, for all $C_\alpha \subseteq W$, there exists $\tau_\alpha > 0$ such that*

i) there exists $\omega > 0$ such that

$$\begin{array}{c} \overset{\circ}{\mathbb{D}}_{\tau_\alpha} \times \partial C_\alpha \;\cap\; \Psi^{-1}\big((\overset{\circ}{\mathbb{D}}_\omega - \mathbf{0}) \times \overset{\circ}{\mathbb{D}}_{\tau_\alpha}\big) \\ \big\downarrow \Psi := (f, t) \\ \big(\overset{\circ}{\mathbb{D}}_\omega - \mathbf{0}\big) \times \overset{\circ}{\mathbb{D}}_{\tau_\alpha} \end{array}$$

is a proper, stratified submersion;

ii) for all δ with $0 < \delta < \tau_\alpha$, there $\xi > 0$ such that

$$\begin{array}{c} \mathbb{D}_\delta \times C_\alpha \;\cap\; f^{-1}(\overset{\circ}{\mathbb{D}}_\xi - \mathbf{0}) \\ \big\downarrow f \\ \overset{\circ}{\mathbb{D}}_\xi - \mathbf{0} \end{array}$$

is a proper, stratified submersion, where the strata are the cross-product strata of $\overset{\circ}{\mathbb{D}}_\delta \times C_\alpha$ together with those of $\partial\mathbb{D}_\delta \times C_\alpha$; and

iii) $\{\mathbb{D}_\delta \times C_\alpha \mid 0 < \delta < \tau_\alpha\}$ is a system of Milnor neighborhoods for f at the origin and hence, by ii), is in fact a system of weakly privileged neighborhoods.

Proof. There exists an open neighborhood of the origin in $\overset{\circ}{\mathbb{D}} \times \mathcal{U}$ of the form $\overset{\circ}{\mathbb{D}}_\eta \times W$ such that $(\overset{\circ}{\mathbb{D}}_\eta \times W) \cap \Sigma f \subseteq V(f)$. As $V(t)$ is prepolar, we may assume that $\mathfrak{G}$ is defined inside $\overset{\circ}{\mathbb{D}}_\eta \times W$ and that $V(t)$ transversely intersects all strata of $\mathfrak{G}$, other than the origin, inside $\overset{\circ}{\mathbb{D}}_\eta \times W$. Finally, as $V(t)$ is prepolar, we may use Theorem 1.28 to conclude that $\gamma^1_{f,t}(\mathbf{0})$ exists and, hence, we may select $\overset{\circ}{\mathbb{D}}_\eta \times W$ so that $(0 \times W) \cap \Gamma^1_{f,t} \subseteq \{\mathbf{0}\}$. Let $C_\alpha \subseteq W$.

i) This follows the proof of Proposition 2.1 of [**Lê1**], applied to each stratum of ∂C_α. Suppose the contrary. Then, we would have a stratum S of C_α and a sequence of points $\mathbf{p}_i$ not in $V(f)$ but in $\mathbb{C} \times S$ such that $\mathbf{p}_i = (t_i, \mathbf{q}_i) \to \mathbf{p} = (0, \mathbf{q}) \in V(t) \cap V(f)$ and such that

$$(*) \qquad T_{\mathbf{p}_i} V(f - f(\mathbf{p}_i), t - t_i) + T_{\mathbf{p}_i}(\mathbb{C} \times S) \neq \mathbb{C}^{n+2}.$$

(That $T_{\mathbf{p}_i} V(f - f(\mathbf{p}_i), t - t_i)$ exists is not completely trivial – it follows from the assumptions made in the preceding paragraph.) Let G denote the good stratum of $V(f)$ containing $\mathbf{p}$. Note that G cannot be the point-stratum $\{\mathbf{0}\}$ as $\mathbf{p}$ is contained in $0 \times \partial C_\alpha$. Let R denote the stratum of ∂C_α containing $\mathbf{q}$.

By taking a subsequence if necessary, we may assume that $T_{\mathbf{p}_i} V(f - f(\mathbf{p}_i))$ converges to some T and that $T_{\mathbf{q}_i} S$ converges to some $\mathcal{T}$. By the Thom condition, $T_{\mathbf{p}} G \subseteq T$ and, by Whitney's condition a), $T_{\mathbf{q}} R \subseteq \mathcal{T}$. Furthermore, as $V(t)$ is prepolar, $V(t)$ transversely intersects G at $\mathbf{p}$.

Thus,

$$T_{\mathbf{p}_i}(\mathbb{C} \times S) \to \mathbb{C} \times \mathcal{T}$$

and

$$T_{\mathbf{p}_i} V(f - f(\mathbf{p}_i), t - t_i) \to T \cap T_{\mathbf{p}} V(t).$$

Also, we have that

$$T_{\mathbf{p}}(G \cap V(t)) = T_{\mathbf{p}} G \cap T_{\mathbf{p}} V(t) \subseteq T \cap T_{\mathbf{p}} V(t)$$

and we know that

$$T_{\mathbf{p}}(G \cap V(t)) + T_{\mathbf{p}}(0 \times S) = 0 \times \mathbb{C}^{n+1},$$

as $\{C_\alpha\}$ is a system of privileged neighborhoods with respect to $\mathfrak{G} \cap V(t)$. It follows at once that

$$T \cap T_{\mathbf{p}} V(t) + \mathbb{C} \times \mathcal{T} = \mathbb{C}^{n+2},$$

but this contradicts (*). This proves i).

ii) That f can be made a submersion on $\mathring{\mathbb{D}}_\delta \times \mathring{C}_\alpha$ follows from the fact that $\mathring{\mathbb{D}}_\delta \times \mathring{C}_\alpha \subseteq \mathring{\mathbb{D}}_\eta \times W$ and $(\mathring{\mathbb{D}}_\eta \times W) \cap \Sigma f \subseteq V(f)$.

That f can be made a stratified submersion on $\mathring{\mathbb{D}}_\delta \times \partial C_\alpha$ and on $\partial \mathbb{D}_\delta \times \partial C_\alpha$ is exactly the argument of i).

Thus, what remains to be shown is that f can be made a submersion on the stratum $\partial \mathbb{D}_\delta \times \mathring{C}_\alpha$. By Theorem 1.28 and Proposition 1.23, $\dim_{\mathbf{0}}(\Gamma^1_{f,t} \cap V(f)) \leqslant 0$ and thus we may assume that $\Gamma^1_{f,t} \cap V(f) \cap (\partial \mathbb{D}_\delta \times C_\alpha)$ is empty.

As $\Gamma^1_{f,t} \cap (\partial \mathbb{D}_\delta \times C_\alpha)$ is compact, $|f|$ obtains a minimum, $\xi > 0$, on $\Gamma^1_{f,t} \cap (\partial \mathbb{D}_\delta \times C_\alpha)$. Now, consider the critical points of f restricted to $\partial \mathbb{D}_\delta \times \mathring{C}_\alpha$ that occur in $f^{-1}(\mathring{\mathbb{D}}_\xi - 0)$. These points occur precisely on

$$\Gamma^1_{f,t} \cap (\partial \mathbb{D}_\delta \times \mathring{C}_\alpha) \cap f^{-1}(\mathring{\mathbb{D}}_\xi - 0)$$

which we know is empty. This proves ii).

iii) We first need two results.

a) for all ω_1, ω_2 with $0 < \omega_1 < \omega_2 < \tau_\alpha$, there exists $\xi > 0$ such that

$$\begin{array}{c} \mathbb{C} \times C_\alpha \ \cap \ \Phi^{-1}((\mathring{\mathbb{D}}_\xi - 0) \times [\omega_1^2, \omega_2^2]) \\ \big\downarrow \Phi := (f, |t|^2) \\ (\mathring{\mathbb{D}}_\xi - 0) \times [\omega_1^2, \omega_2^2] \end{array}$$

is a proper, stratified submersion and thus, for all $\eta \in \mathring{\mathbb{D}}_\xi - 0$, the inclusion

$$(\mathbb{D}_{\omega_1} \times C_\alpha) \cap V(f - \eta) \hookrightarrow (\mathbb{D}_{\omega_2} \times C_\alpha) \cap V(f - \eta)$$

is a homotopy-equivalence; and

b) if $C_\alpha \subseteq \mathring{C}_\beta$, then there exist $\tau, \xi > 0$ such that, for all $\delta \in \mathring{\mathbb{D}}_\tau - 0$ and $\eta \in \mathring{\mathbb{D}}_\xi - 0$, the inclusion

$$(\mathbb{D}_\delta \times C_\alpha) \cap V(f - \eta) \hookrightarrow (\mathbb{D}_\delta \times C_\beta) \cap V(f - \eta)$$

is a homotopy-equivalence.

Assuming a) and b) for the moment, we proceed with the proof. Suppose that $\mathbb{D}_\sigma \times C_\alpha \subseteq \mathbb{D}_\rho \times \mathring{C}_\beta$.

By b), for all small, non-zero δ and η,

$$(\mathbb{D}_\delta \times C_\alpha) \cap V(f - \eta) \hookrightarrow (\mathbb{D}_\delta \times C_\beta) \cap V(f - \eta)$$

is a homotopy-equivalence. If we select δ so small that $\mathbb{D}_\delta$ is contained in both $\mathbb{D}_\sigma$ and $\mathbb{D}_\rho$, then we may apply a) twice to obtain that, for all small, non-zero η,

$$(\mathbb{D}_\delta \times C_\alpha) \cap V(f - \eta) \hookrightarrow (\mathbb{D}_\sigma \times C_\alpha) \cap V(f - \eta)$$

and

$$(\mathbb{D}_\delta \times C_\beta) \cap V(f - \eta) \hookrightarrow (\mathbb{D}_\sigma \times C_\beta) \cap V(f - \eta)$$

are homotopy-equivalences. The conclusion that

$$(\mathbb{D}_\delta \times C_\alpha) \cap V(f - \eta) \hookrightarrow (\mathbb{D}_\rho \times C_\beta) \cap V(f - \eta)$$

is a homotopy-equivalence now follows immediately by combining the three previous homotopy-equivalences.

We now prove a) and b).

Proof of a): That Φ is a stratified submersion on $\mathbb{C}\times\partial C_\alpha$ is once again exactly the proof of i). That Φ is a submersion on $\mathbb{C}\times\overset{\circ}{C}_\alpha$ is similar to our argument in ii): as $\dim_{\mathbf{0}}(\Gamma^1_{f,t}\cap V(f))\leqslant 0$, we may assume that

$$\Gamma^1_{f,t}\cap V(f)\cap((\mathbb{D}_{\omega_2}-\overset{\circ}{\mathbb{D}}_{\omega_1})\times C_\alpha)$$

is empty. Therefore, by compactness, $|f|$ obtains a minimum, $\xi>0$, on $\Gamma^1_{f,t}\cap((\mathbb{D}_{\omega_2}-\overset{\circ}{\mathbb{D}}_{\omega_1})\times C_\alpha)$. Now, consider the critical points of Φ restricted to $\mathbb{C}\times\overset{\circ}{C}_\alpha$ that occur in $\Phi^{-1}((\overset{\circ}{\mathbb{D}}_\xi-0)\times[\omega_1^2,\omega_2^2])$. These points occur precisely in

$$\Gamma^1_{f,t}\cap f^{-1}(\overset{\circ}{\mathbb{D}}_\xi-0)\cap((\mathbb{D}_{\omega_2}-\overset{\circ}{\mathbb{D}}_{\omega_1})\times C_\alpha)$$

which we know is empty. This proves a).

Proof of b): Let $C_\alpha\subseteq\overset{\circ}{C}_\beta$. Let τ be so small that inside $\mathbb{D}_\tau\times C_\beta$ all points of $\Gamma^1_{f,t}$ occur in $\mathbb{D}_\tau\times\overset{\circ}{C}_\alpha$. Further, choose $\tau<\min\{\tau_\alpha,\tau_\beta\}$ so that we may apply i) in both cases. Choose ξ so small that $(C_\alpha,\mathbb{D}_\xi)$ and $(C_\beta,\mathbb{D}_\xi)$ are Milnor pairs for $f_{|V(t)}$ and so small that we may apply i) to both $\mathbb{D}_\tau\times\partial C_\alpha$ and $\mathbb{D}_\tau\times\partial C_\beta$ over $(\overset{\circ}{\mathbb{D}}_\xi-0)\times\overset{\circ}{\mathbb{D}}_\tau$. Fix some $\delta\in\overset{\circ}{\mathbb{D}}_\tau-0$ and $\eta\in\overset{\circ}{\mathbb{D}}_\xi-0$.

By i) or ii), $V(f-\eta)$ transversely intersects all the strata of $\mathbb{C}\times\partial C_\alpha$ and $\mathbb{C}\times\partial C_\beta$, so may Whitney stratify $(\mathbb{C}\times C_\beta)\cap V(f-\eta)$ by taking as strata the intersection of $V(f-\eta)$ with each of $\mathbb{C}\times\overset{\circ}{C}_\alpha$, $\mathbb{C}\times(\overset{\circ}{C}_\beta-C_\alpha)$, and the strata of $\mathbb{C}\times\partial C_\alpha$ and $\mathbb{C}\times\partial C_\beta$.

As $C_\alpha\cap V(f_{|V(t)}-\eta)\hookrightarrow C_\beta\cap V(f_{|V(t)}-\eta)$ is a homotopy-equivalence and $V(f_{|V(t)})-\eta)$ transversely intersects ∂C_α and ∂C_β, for all small $\mathbb{D}_\mu$ we must have that

$$(\mathbb{D}_\mu\times C_\alpha)\cap V(f-\eta)\hookrightarrow(\mathbb{D}_\mu\times C_\beta)\cap V(f-\eta)$$

is also a homotopy-equivalence. We wish to pass from $\mathbb{D}_\mu$ to $\mathbb{D}_\delta$ by considering the function $|t|^2$ on the stratified space $(\mathbb{C}\times C_\beta)\cap V(f-\eta)$ (with the stratification given above).

By i), $|t|^2$ has no critical points on the strata of $(\mathbb{C}\times\partial C_\alpha)\cap V(f-\eta)$ and $(\mathbb{C}\times\partial C_\beta)\cap V(f-\eta)$ when $|t|<\delta$. In addition, the critical points on the interior strata, $(\mathbb{C}\times\overset{\circ}{C}_\alpha)\cap V(f-\eta)$ and $(\mathbb{C}\times(\overset{\circ}{C}_\beta-C_\alpha))\cap V(f-\eta)$, occur on the polar curve and, hence, by our earlier requirement, these critical points all occur in $\mathbb{C}\times\overset{\circ}{C}_\alpha$. Therefore, using stratified Morse theory **[G-M2]** together with the homotopy-equivalence lemma 3.7 of **[Mi2]**, we find that the inclusion

$$(\mathbb{D}_\delta\times C_\alpha)\cap V(f-\eta)\hookrightarrow(\mathbb{D}_\delta\times C_\beta)\cap V(f-\eta)$$

is a homotopy-equivalence. □

For a family of analytic functions $f_t : (\mathcal{U}, \mathbf{0}) \to (\mathbb{C}, 0)$, we are interested in how the Milnor fibre and fibration "jump" as we move from small non-zero t to $t = 0$. Hence, we make the following definition.

Definition A.7. If we are considering the family $f_t : (\mathcal{U}, \mathbf{0}) \to (\mathbb{C}, 0)$, we refer to i) of A.6 by saying that the family satisfies the *conormal condition with respect to* $\{C_\alpha\}$.

The point of this condition is that it says that the Milnor fibration of f_0 lifts trivially in the family f_t on the boundary of the neighborhoods C_α.

Definition A.8. The *Thom set at the origin*, $\mathfrak{T}_f$, is the set of $(n+1)$-planes which occur as limits at the origin of the tangent spaces to level hypersurfaces of f, i.e. $T \in \mathfrak{T}_f$ if and only if there exists a sequence of points $\mathbf{p}_i$ in $\mathring{\mathbb{D}} \times \mathcal{U} - \Sigma f$ such that $\mathbf{p}_i \to \mathbf{0}$ and $T = \lim T_{\mathbf{p}_i} V(f - f(\mathbf{p}_i))$.

Equivalently, $\mathfrak{T}_f$ is the fibre over the origin in the Jacobian blow-up of f (see [**H-L**]). $\mathfrak{T}_f$ is thus a closed algebraic subset of the Grassmanian $G_{n+1}(\mathbb{C}^{n+2}) =$ the projective space of $(n+1)$-planes in $\mathbb{C}^{n+2}$.

Proposition A.9. *Suppose that* $V(t)$ *is a prepolar slice for* f *at* $\mathbf{0}$ *or that* $V(t) = T_{\mathbf{0}} V(t) \notin \mathfrak{T}_f$. *Then,*

i) $\dim_{\mathbf{0}} \Gamma^1_{f,t} \leqslant 1$, *and*

ii) the family f_t *satisfies the conormal condition with respect to any universal system of privileged neighborhoods,* $\mathcal{C}$, *for* f_0 *at* $\mathbf{0}$.

Moreover, whenever i) and ii) are satisfied, there is an inclusion of the Milnor fibre $F_{f_{t_0},\mathbf{0}}$ *into the Milnor fibre* $F_{f_0,\mathbf{0}}$ *for all small non-zero* t_0; *the homotopy-type of this inclusion is independent of the choice of* t_0 *and the choice of the universal system of privileged neighborhoods,* $\mathcal{C}$.

Proof. That there is such an inclusion whenever i) and ii) are satisfied is standard. One considers the map $\Psi := (f, t)$ and its restriction

$$\begin{array}{c}(\mathbb{D}_\tau \times C) \cap \Psi^{-1}\left((\mathbb{D}_\xi - 0) \times \mathbb{D}_\tau\right)\\ \downarrow \Psi \\ (\mathbb{D}_\xi - 0) \times \mathbb{D}_\tau\end{array}$$

for appropriately small choices of $C \in \mathcal{C}$, ξ, and τ. By the conormal condition, this is a stratified submersion on the boundary. As $\dim_{\mathbf{0}} \Gamma^1_{f,t} \leqslant 1$, the discriminant of Ψ, $\Psi(\Gamma^1_{f,t})$, is also at most one-dimensional. Thus, we may lift a path in the base which avoids the discriminant to get a diffeomorphism between the Milnor fibre of f_0 and $C \cap V(f_{t_0} - \eta)$ for all small t_0 and for all η with

$0 < |\eta| \ll |t_0|$. And, though we do not know that $\mathcal{C}$ is a system of privileged neighborhoods for f_{t_0}, we may still take a small enough ball inside C to obtain the desired inclusion, which is clearly independent of the choice of t_0.

That the inclusion is independent of the choice of privileged neighborhoods follows similarly. Suppose that $\mathcal{C}'$ is second universal system of privileged neighborhoods for f_0. Let $C \in \mathcal{C}$ and let $C' \in \mathcal{C}'$ be such that $C' \subseteq \overset{\circ}{C}$, and such that C and C' are small enough to give the Milnor fibre, i.e. for all small non-zero ξ, the inclusion of $C' \cap V(f_0 - \xi)$ into $C \cap V(f_0 - \xi)$ is a homotopy-equivalence where both spaces are homotopy-equivalent to the Milnor fibre of f_0 at the origin. Then, as above, over a curve which avoids the discriminant, we have a proper, stratified submersion – where the strata are those of $\mathbb{D}_\tau \times \partial C'$ together with those of $\mathbb{D}_\tau \times \partial C'$ plus the interior.

Hence, the homotopy-equivalence $C' \cap V(f_0 - \xi) \to C \cap V(f_0 - \xi)$ lifts to a homotopy-equivalence $C' \cap V(f_{t_0} - \xi) \to C \cap V(f_{t_0} - \xi)$. The independence statement now follows easily.

We must still show that if $V(t)$ is a prepolar slice for f at $\mathbf{0}$ or $V(t) = T_{\mathbf{0}}V(t) \notin \mathfrak{T}_f$, then i) and ii) hold.

If $V(t)$ is prepolar for f at $\mathbf{0}$, then i) follows from Theorem 1.28 and ii) follows from A.6.i. If $V(t) \notin \mathfrak{T}_f$, then clearly $\Gamma^1_{f,t}$ is empty near the origin. It remains for us to show that if $V(t) \notin \mathfrak{T}_f$, then the family f_t satisfies the conormal condition with respect to any universal system of privileged neighborhoods, $\mathcal{C}$, for f_0.

If $V(t) \notin \mathfrak{T}_f$, then $V(t)$ certainly transversely intersects the smooth part of $V(f)$ in a neighborhood of the origin. Hence, we may use Proposition A.2 to conclude that there exists a good stratification, $\mathfrak{G}$, for f at the origin such that the strata of $\mathfrak{G}$ which are contained in $V(t)$ form a good stratification for f_0 at the origin. The proof now proceeds like that of A.6.i.

Suppose to the contrary that, for arbitrarily small C_α in $\mathcal{C}$, there exists a stratum S of ∂C_α and a sequence of points $\mathbf{p}_i$ not in $V(f)$ but which are in $\mathbb{C} \times S$ such that $\mathbf{p}_i = (t_i, \mathbf{q}_i) \to \mathbf{p} := (0, \mathbf{q}) \in V(t) \cap V(f)$ and such that

$$(*) \qquad T_{\mathbf{p}_i}V(f - f(\mathbf{p}_i), t - t_i) + T_{\mathbf{p}_i}(\mathbb{C} \times S) \neq \mathbb{C}^{n+2}.$$

Let G denote the good stratum of $V(f)$ which contains $\mathbf{p}$. Note that G is contained in $V(t)$ by the nature of our good stratification and that G cannot be simply the stratum consisting of the origin since $\mathbf{p}$ is contained in $0 \times \partial C_\alpha$. Let R denote the stratum of ∂C_α containing $\mathbf{q}$.

By taking a subsequence if necessary, we may assume that $T_{\mathbf{p}_i}V(f - f(\mathbf{p}_i))$ converges to some T and that $T_{\mathbf{q}_i}S$ converges to some $\mathcal{T}$. By the Thom condition, $T_{\mathbf{p}}G \subseteq T$ and, by Whitney's condition a), $T_{\mathbf{q}}R \subseteq \mathcal{T}$. Furthermore, as $V(t) \notin \mathfrak{T}_f$, we may assume that $\mathbf{p}$ is close enough to the origin that $T \neq V(t)$.

Thus, $T_{\mathbf{p}_i}(\mathbb{C} \times S) \to \mathbb{C} \times \mathcal{T}$ and $T_{\mathbf{p}_i}V(f - f(\mathbf{p}_i), t - t - i) \to T \cap T_{\mathbf{p}_i}V(t)$. Also, we have that $T_{\mathbf{p}}G = T_{\mathbf{p}}G \cap T_{\mathbf{p}}V(t) \subseteq T \cap T_{\mathbf{p}}V(t)$, and we know that $T_{\mathbf{p}}G + T_{\mathbf{p}}(0 \times S) = 0 \times \mathbb{C}^{n+1}$, as $\{C_\alpha\}$ is a system of privileged neighborhoods with respect to $\mathfrak{G} \cap V(t)$. It follows at once that $T \cap T_{\mathbf{p}}V(t) + \mathbb{C} \times T = \mathbb{C}^{n+2}$ – which contradicts $(*)$. □

If the polar curve, $\Gamma^1_{f,t}$, is empty, then the map Ψ which appears in the proof of Proposition A.9 is a stratified submersion over the entire base space and so, for all small $t_0 \neq 0$, we have a fibre-prerserving inclusion of the total space of the Milnor fibration of f_{t_0} into the total space of the Milnor fibration of f_0. Moreover, exactly as above, this inclusion is independent – up to homotopy – of all of the choices made. By the theorem of Dold (see [**Hu**, p. 209]), this inclusion is a fibre homotopy-equivalence if and only if the inclusion of each fibre is a homotopy-equivalence. Therefore, we make the following definitions.

Definition A.10 Whenever i) and ii) of A.9 hold, we say that the family, f_t, satisfies the *universal conormal condition*.

If f_t satisfies the universal conormal condition, we say that f_t has the *homotopy Milnor fibre lifting property* if and only if the inclusion of A.9 is a homotopy-equivalence.

If f_t satisfies the universal conormal condition, we say that f_t has the *homology Milnor fibre lifting property* if and only if the inclusion of A.9 induces isomorphisms on all integral homology groups.

The family, f_t, has the *homotopy Milnor fibration lifting property* if and only if f_t has the homotopy Milnor fibre lifting property and $\Gamma^1_{f,t} = \emptyset$ in a neighborhood of the origin. This definition makes sense in light of our above discussion concerning the result of Dold.

One may also discuss the Milnor fibre and Milnor fibration up to diffeomorphism if one is willing to restrict consideration to the standard universal system of Milnor neighborhoods, namely the set of closed balls centered at the origin. In this case, we may use the h-cobordism Theorem and the pseudo-isotopy result of Cerf [**Ce**] to translate the homotopy information into smooth information – provided that we are in a sufficiently high dimension and that the Milnor fibre and its boundary are sufficiently connected. More specifically, if $\mathcal{U}$ is an open neighborhood of the origin in $\mathbb{C}^{n+1}$, $f_t : (\mathcal{U}, \mathbf{0}) \rightarrow (\mathbb{C}, 0)$ has the homotopy Milnor fibration lifting property, $n \geqslant 3$, and the Milnor fibre and its boundary are simply-connected for each f_t for all small t, then the diffeomorphism-type of the Milnor fibrations is constant in the family near $t = 0$. This connectedness condition can be realized by requiring $n - \dim_{\mathbf{0}}\Sigma f_0 \geqslant 3$ (see [**K-M**] and [**Ra**]).

We wish to state the diffeomorphism results discussed above precisely. First, we give without proof Cerf's pseudo-isotopy result in the form that we shall need it.

Lemma A.11. *Let X be a smooth manifold with boundary $\partial X = X_0 \dot{\cup} X_1$*

and let $\pi : X \to S^1$ *be a smooth locally trivial fibration over a circle with fibre diffeomorphic to* $M \times [0,1]$, *where* M *is a closed, simply-connected, smooth manifold of dimension* $\geqslant 5$.

Then, the restriction of π *to* X_0 *is a smooth locally trivial fibration with fibre diffeomorphic to* M, *and there exists a commutative diagram*

$$\begin{array}{ccc} (X, X_0) & \xrightarrow[\text{diffeo.}]{\cong} & (X_0 \times [0,1], X_0 \times \{0\}) \\ & \pi \searrow \quad \swarrow \pi_{|X_0} \circ pr_1 & \\ & S^1 & \end{array}$$

where the diffeomorphism is the identity on $X_0 = X_0 \times \{0\}$.

Proposition A.12. *Let* $\mathcal{U}$ *be an open neighborhood of the origin in* $\mathbb{C}^{n+1}$. *Suppose that the family* $f_t : (\mathcal{U}, \mathbf{0}) \to (\mathbb{C}, 0)$ *has the homotopy Milnor fibration lifting property and* $n - \dim_{\mathbf{0}} \Sigma f_0 \geqslant 3$. *Then, the diffeomorphism-type of the Milnor fibrations of* f_t *at the origin is independent of* t *for all small* t.

Proof. We shall use the notation from the proof of Proposition A.9. We fix the universal system of privileged neighborhoods to be the collection of closed balls centered at the origin.

As f_t has the homotopy Milnor fibration lifting property, the polar curve $\Gamma^1_{f,t}$ is empty and so the map Ψ in the proof of Proposition A.9 is a proper stratified submersion. Hence, for $0 < \xi, |t_0| \ll \epsilon$, the Milnor fibration of f_0 is diffeomorphic to $B_\epsilon \cap f_{t_0}^{-1}(\partial \mathbb{D}_\xi) \xrightarrow{f_{t_0}} \partial \mathbb{D}_\xi$. The problem, of course, is that B_ϵ may be too large a ball in which to define the Milnor fibration of f_{t_0}. Let F and E denote the fibre and the total space, respectively, of this previous fibration.

Let F' denote the Milnor fibre of f_{t_0} at the origin, where we again use closed balls for the Milnor neighborhoods. Let E' denote the total space of the Milnor fibration f_{t_0}.

As f_t has the homotopy Milnor fibration lifting property, the inclusion of E' into E induces an inclusion $F' \hookrightarrow F$ which is a homotopy-equivalence.

Since $n - \dim_{\mathbf{0}} \Sigma f_0 \geqslant 3$, F, F', ∂F, and $\partial F'$ are simply-connected (see [**Ra**]). Combining this with the fact that $F' \hookrightarrow F$ is a homotopy-equivalence, we may duplicate the argument of Lê and Ramanujam [**L-R**] to conclude that $\Delta T := E - \overset{\circ}{E'}$ is the total space of a differentiable fibration over $\partial \mathbb{D}_\xi$ with projection f_{t_0} and fibre $F - \overset{\circ}{F'}$ which is diffeomorphic to $\partial F \times [0,1]$ via the h-cobordism theorem.

Now, by Lemma A.11, $\Delta T \xrightarrow{f_{t_0}} \partial \mathbb{D}_\xi$ is diffeomorphic to

$$\partial E' \times [0,1] \xrightarrow{f_{t_0} \circ pr_1} \partial \mathbb{D}_\xi \times \{0\}$$

by a diffeomorphism which is the identity on $\partial E' = \partial E' \times \{0\}$. Combining this with a fibred collar of $\partial E'$ in E', we conclude that $E' \xrightarrow{f_{t_0}} \partial\mathbb{D}_\xi$ is diffeomorphic to $E \xrightarrow{f_{t_0}} \partial\mathbb{D}_\xi$, which we already know is diffeomorphic to the Milnor fibration of f_0 at $\mathbf{0}$. □

We now wish to prove a fundamental result – namely, that if we have a family f_t in which the Milnor fibrations of a hyperplane slice are independent of t and the number of handles attached in passing from the Milnor fibre of the hyperplane slice to the entire Milnor fibre is constant, then the Milnor fibrations are constant in the family. Despite the fact that the dimension of the critical loci is allowed to be arbitrary, the argument is exactly that which we used in [**Mas2**] where the critical loci were all one-dimensional.

Theorem A.13. *Let $\mathcal{W}$ be an open neighborhood of the origin in $\mathbb{C}^{n+2}$ and let $g_t : (\mathcal{W}, \mathbf{0}) \to (\mathbb{C}, 0)$ be an analytic family. Let s denote* $\dim_{\mathbf{0}}\Sigma g_0$. *Assume that g_t satisfies the universal conormal condition and that L is a linear form such that $V(L)$ is prepolar for f_t at the origin for all small t and such that $g_{t|_{V(L)}}$ satisfies the universal conormal condition. Suppose further that $\left(\Gamma^1_{g_t,L} \cdot V(g_t)\right)_{\mathbf{0}}$ is constant for all small t.*

Under the above assumptions, if $g_{t|_{V(L)}}$ has the homology Milnor fibre lifting property, then g_t has the homology Milnor fibre lifting property.

Moreover, if $s \leqslant n-1$ and $g_{t|_{V(L)}}$ has the homotopy Milnor fibre lifting property, then g_t has the homotopy Milnor fibre lifting property.

Proof. This is actually quite trivial. Let F_0 and F_{t_0} denote the Milnor fibre of g_0 and g_{t_0} for small non-zero t_0, respectively. The Milnor fibres of $g_{0|_{V(L)}}$ and $g_{t_0|_{V(L)}}$ are then $F_0 \cap V(L)$ and $F_{t_0} \cap V(L)$, respectively. Let γ denote the constant value of $\left(\Gamma^1_{g_t,L} \cdot V(g_t)\right)_{\mathbf{0}}$.

As g_t and $g_{t|_{V(L)}}$ satisfy the universal conormal condition, we may repeat the argument of Proposition A.9 – lifting a path in the base which avoids the discriminants of both (g_t, t) and $(g_{t|_{V(L)}}, t)$ – to obtain compatible inclusions $F_{t_0} \hookrightarrow F_0$ and $F_{t_0} \cap V(L) \hookrightarrow F_0 \cap V(L)$.

Suppose that $g_{t|_{V(L)}}$ has the homology Milnor fibre lifting property, i.e. $F_{t_0} \cap V(L) \hookrightarrow F_0 \cap V(L)$ induces isomorphisms on homology. We wish to show that $F_{t_0} \hookrightarrow F_0$ induces isomorphisms on homology. We will accomplish this by showing that $H_*(F_0, F_{t_0}) = 0$.

By considering the homology long exact sequence of the triple $(F_0, F_0 \cap V(L), F_{t_0} \cap V(L))$, we find that $H_i(F_0, F_{t_0} \cap V(L)) \cong H_i(F_0, F_0 \cap V(L))$ for all i. By Lê's attaching result (Theorem 0.9) or Theorem 3.1, $H_i(F_0, F_0 \cap V(L)) = 0$ unless $i = n+1$ and $H_{n+1}(F_0, F_0 \cap V(L)) \cong \mathbb{Z}^\gamma$.

Now, we are going to consider the homology long exact sequence of the triple $(F_0, F_{t_0}, F_{t_0} \cap V(L))$. From the last paragraph, we know that $H_i(F_0, F_{t_0} \cap$

$V(L)) = 0$ unless $i = n+1$. In addition, $H_i(F_{t_0}, F_{t_0} \cap V(L)) = 0$ unless $i = n+1$. Moreover,

$$H_{n+1}(F_0, F_0 \cap V(L)) \cong H_{n+1}(F_{t_0}, F_{t_0} \cap V(L)) \cong \mathbb{Z}^\gamma.$$

Thus, in the long exact sequence of the triple $(F_0, F_{t_0}, F_{t_0} \cap V(L))$, all terms are zero except in the portion

$$0 \to \mathbb{Z}^\gamma \to \mathbb{Z}^\gamma \to H_{n+1}(F_0, F_{t_0}) \to 0.$$

But, as in the proof of the result of Lê and Ramanujam **[L-R]**, $H_{n+1}(F_0, F_{t_0})$ is free Abelian, since F_0 is obtained from F_{t_0} by attaching handles of index less than or equal to $n+1$. (One considers the function distance squared from the origin and lets the function grow from the small ball used to define F_{t_0} out to the ball used to define F_0. One hits no critical points of index greater than or equal to $n+2$.) Thus, $H_{n+1}(F_0, F_{t_0}) = 0$ and we have proved the first claim.

The second claim follows from the first, since $s \leqslant n-1$ guarantees that F_0 and F_{t_0} are simply-connected, and then we apply the Whitehead Theorem. □

There are two more big results which we need to prove in this appendix – both deal with suspending singularities (see Chapter 8). The first result is that there exists a universal system of privileged neighborhoods of a particularly nice form for the function $h + w^j$, where w is a variable disjoint from those of h. The second result says, with a few extra assumptions, that the constancy of the Milnor fibrations in the family f_t implies the constancy of the Milnor fibrations in the family $f_t + w^j$, where, again, w is disjoint from the variables of f_t. This second result seems reasonable since the result of Proposition 8.1 is that the Milnor fibre of $f_t + w^j$ at the origin is homotopy-equivalent to one-point union of $j-1$ copies of the Milnor fibre of f_t at the origin. However, both of these results are technical nightmares.

Let $\mathcal{U}$ be an open neighborhood of the origin in $\mathbb{C}^{n+1}$ and let $h : (\mathcal{U}, \mathbf{0}) \to (\mathbb{C}, 0)$ be an analytic function. Let $j \geqslant 2$ and define $\tilde{h}(w, \mathbf{z}) := h(\mathbf{z}) + w^j$. We wish to show that the set $\{\mathbb{D}_\omega \times B_\epsilon^{2n+2} \mid 0 < \omega \ll \epsilon\}$ is a universal system of privileged neighborhoods for $\tilde{h}$ at the origin. Note that we may **not** use A.6 to conclude that $\{\mathbb{D}_\omega \times B_\epsilon^{2n+2} \mid 0 < \omega \ll \epsilon\}$ is even weakly privileged since the slice $V(w)$ contains the entire critical locus of $h + w^j$ and, hence, is certainly not prepolar for $h + w^j$. Of course, the actual argument is very similar to the proof of Proposition A.6.

Proposition A.14. *The set $\{\mathbb{D}_\omega \times B_\epsilon^{2n+2} \mid 0 < \omega \ll \epsilon\}$ is a universal system of privileged neighborhoods for $h + w^j$ at the origin.*

Proof. As $j \geqslant 2$, $\Sigma(h + w^j) = \{0\} \times \Sigma h$. Fix any good stratification, $\mathfrak{S}$, for $h + w^j$ at the origin in $\mathbb{C}^{n+2}$.

Let $\epsilon_0 > 0$ be so small that the critical locus of the map h inside $B^{2n+2}_{\epsilon_0}$ is contained in $V(h)$ and so small that, for all ϵ with $0 < \epsilon \leqslant \epsilon_0$,

$\{0\} \times \partial B^{2n+2}_\epsilon$ transversely intersects all strata of $\{0\} \times \Sigma V(h)$ inside $\{0\} \times \mathbb{C}^{n+1}$,

$\partial B^{2n+4}_\epsilon$ transversely intersects all strata of $\mathfrak{S}$ (we write $\partial B^{2n+4}_\epsilon \pitchfork \mathfrak{S}$), and

$\partial B^{2n+2}_\epsilon$ transversely intersects all strata of some good stratification for h at $\mathbf{0}$.

This last condition guarantees, for all small non-zero ζ, that

$$(*) \qquad \partial B^{2n+2}_\epsilon \pitchfork V(h - h(\zeta)).$$

Now fix an ϵ between 0 and ϵ_0. We wish to show that there exists $\omega_\epsilon > 0$ such that, for all ω with $0 < \omega \leqslant \omega_\epsilon$, we have:

a) $\overset{\circ}{\mathbb{D}}_\omega \times \partial B^{2n+2}_\epsilon \pitchfork \mathfrak{S}$;

b) $\partial \mathbb{D}_\omega \times \overset{\circ}{B}{}^{2n+2}_\epsilon \pitchfork \mathfrak{S}$;

c) $\partial \mathbb{D}_\omega \times \partial B^{2n+2}_\epsilon \pitchfork \mathfrak{S}$.

After we show this, it will still remain to show that, if $\mathbb{D}_{\omega_1} \times B_{\epsilon_1} \subseteq \mathbb{D}_{\omega_2} \times B_{\epsilon_2}$, then, for all small non-zero t, the inclusion

$$\mathbb{D}_{\omega_1} \times B_{\epsilon_1} \cap V(h + w^j - t) \hookrightarrow \mathbb{D}_{\omega_2} \times B_{\epsilon_2} \cap V(h + w^j - t)$$

is a homotopy-equivalence.

Proof of a): Clearly, as $\{0\} \times \partial B^{2n+2}_\epsilon$ transversely intersects all strata of $\{0\} \times \Sigma V(h)$ inside $\{0\} \times \mathbb{C}^{n+1}$, $\mathbb{C} \times \partial B^{2n+2}_\epsilon$ transversely intersects all singular strata of $V(h + w^j)$. Suppose, however, that no matter how small we pick $\omega > 0$, we still have a point in the smooth stratum, $S := V(h + w^j) - \{0\} \times \Sigma V(h)$, where S does not transversely intersect $\overset{\circ}{\mathbb{D}}_\omega \times \partial B^{2n+2}_\epsilon$.

Then, we would have a sequence $\mathbf{p}_i := (w_i, \mathbf{q}_i) \in \mathbb{C} \times \partial B_\epsilon$ contained in S such that $\mathbf{p}_i \to \mathbf{p} := (0, \mathbf{q}) \in \{0\} \times \partial B_\epsilon$, $T_{\mathbf{p}_i} S \subseteq T_{\mathbf{p}_i}(\mathbb{C} \times \partial B_\epsilon)$, $T_{\mathbf{p}_i} S$ converges to some T, and $T_{\mathbf{p}_i}(\mathbb{C} \times \partial B_\epsilon) \to \mathbb{C} \times T_{\mathbf{q}}(\partial B_\epsilon)$. Let S' denote the stratum of $\mathfrak{S}$ containing $\mathbf{p}$.

By the Thom condition, $T_{\mathbf{p}} S' \subseteq T$ (this is true because T comes from the smooth stratum – we are not assuming Whitney conditions hold between the strata). Hence,

$$T_{\mathbf{p}} S' \subseteq T \subseteq \mathbb{C} \times T_{\mathbf{q}}(\partial B_\epsilon) = T_{\mathbf{p}}(\partial B^{2n+4}_\epsilon),$$

where this last equality is true because the w-coordinate of $\mathbf{p}$ is 0. But, this contradicts the fact that $\partial B^{2n+4}_\epsilon \pitchfork \mathfrak{S}$. This proves a).

Before we prove b) and c), note that if $w \in \partial\mathbb{D}_\omega$, then $w \neq 0$ and, hence, the only stratum of $\mathfrak{S}$ which $\partial\mathbb{D}_\omega \times \overset{\circ}{B}{}^{2n+2}_\epsilon$ and $\partial\mathbb{D}_\omega \times \partial B^{2n+2}_\epsilon$ intersect is the smooth stratum $S := V(h+w^j) - \{0\} \times \Sigma V(h)$.

Proof of b): Actually, we show, regardless of the size of $\omega > 0$, that $\partial\mathbb{D}_\omega \times \overset{\circ}{B}{}^{2n+2}_\epsilon \pitchfork S$.

For if not, we would have $\mathbf{p} := (w, \mathbf{q}) \in S$ such that $w \neq 0$ and $T_{\mathbf{p}}V(h+w^j) \subseteq T_{\mathbf{p}}(\partial\mathbb{D}_{|w|} \times \overset{\circ}{B}{}^{2n+2}_\epsilon)$. This implies that

$$\frac{\partial h}{\partial z_0}_{|\mathbf{q}} = \cdots = \frac{\partial h}{\partial z_n}_{|\mathbf{q}} = 0,$$

i.e. that $\mathbf{q} \in \Sigma h$. Recalling that we chose ϵ such that $B_\epsilon \cap \Sigma h \subseteq V(h)$, we see that $h(\mathbf{q}) = 0$. However, this contradicts that $h(\mathbf{q}) = -w^j \neq 0$. This proves b).

Proof of c): Suppose not. Then, we would have a sequence $\mathbf{p}_i := (w_i, \mathbf{q}_i) \in S \cap (\mathbb{C} \times \partial B^{2n+2}_\epsilon)$ with $w_i \neq 0$, $\mathbf{p}_i \to \mathbf{p} = (0, \mathbf{q}) \in \{0\} \times \partial B_\epsilon$, and such that

$$T_{\mathbf{p}_i}V(h+w^j) + T_{\mathbf{p}_i}(\partial\mathbb{D}_{|w_i|} \times \partial B_\epsilon) \neq \mathbb{C}^{n+2}.$$

This implies that $T_{\mathbf{q}_i}V(h - h(\mathbf{q}_i)) \subseteq T_{\mathbf{q}_i}(\partial B^{2n+2}_\epsilon)$, while $h(\mathbf{q}_i) = -w_i^j$ approaches - but is unequal to - zero. This, however, is impossible by $(*)$. This proves c).

We must still prove the homotopy-equivalence statement. In a manner completely similar to the proofs of a), b), and c) above, one can easily show, using the Thom condition, that the following statements are true:

d) for all ϵ with $0 < \epsilon \leqslant \epsilon_0$, if ω_1 is between 0 and ω_ϵ, then for all ω_2 with $0 < \omega_2 \leqslant \omega_1$, there exists $\xi > 0$ such that

$$\begin{array}{c} (\mathbb{C} \times B^{2n+2}_\epsilon) \cap \Psi^{-1}((\mathbb{D}_\xi - 0) \times [\omega_2^2, \omega_1^2]) \\ \Big\downarrow \ \Psi := (h + w^j, |w|^2) \\ (\mathbb{D}_\xi - 0) \times [\omega_2^2, \omega_1^2] \end{array}$$

is a proper, stratified submersion and therefore

d′) $(\mathbb{D}_{\omega_2} \times B_\epsilon) \cap (h+w^j)^{-1}(\mathbb{D}_\xi - 0) \hookrightarrow (\mathbb{D}_{\omega_1} \times B_\epsilon) \cap (h+w^j)^{-1}(\mathbb{D}_\xi - 0)$

is a fibre-homotopy equivalence between total spaces (where the projection in each case is the obvious map $h + w^j$).

e) if $0 < \epsilon_2 < \epsilon_1 \leqslant \epsilon_0$, then for all small, non-zero ω, there exists $\xi > 0$ such that

$$\begin{array}{c}(\mathbb{D}_\omega \times \mathbb{C}^{n+1}) \cap \Phi^{-1}((\mathbb{D}_\xi - 0) \times [\epsilon_2^2, \epsilon_1^2]) \\ \Big\downarrow \Phi := (h + w^j, |\mathbf{z}|^2) \\ (\mathbb{D}_\xi - 0) \times [\epsilon_2^2, \epsilon_1^2]\end{array}$$

is a proper, stratified submersion and therefore

e′) $\quad (\mathbb{D}_\omega \times B_{\epsilon_2}) \cap (h + w^j)^{-1}(\mathbb{D}_\xi - 0) \hookrightarrow (\mathbb{D}_\omega \times B_{\epsilon_1}) \cap (h + w^j)^{-1}(\mathbb{D}_\xi - 0)$

is a fibre-homotopy equivalence.

Now, suppose that we have $\mathbb{D}_{\omega_1} \times B_{\epsilon_1} \subseteq \mathbb{D}_{\omega_1} \times B_{\epsilon_1}$, where $0 < \epsilon_2 < \epsilon_1 \leqslant \epsilon_0$, $0 < \omega_2 < \omega_1 \leqslant \omega_{\epsilon_1}$, and $\omega_2 < \omega_{\epsilon_2}$. We shall show that, for all small $\xi > 0$,

$$(\mathbb{D}_{\omega_2} \times B_{\epsilon_2}) \cap (h + w^j)^{-1}(\mathbb{D}_\xi - 0) \hookrightarrow (\mathbb{D}_{\omega_1} \times B_{\epsilon_1}) \cap (h + w^j)^{-1}(\mathbb{D}_\xi - 0)$$

is a fibre-homotopy equivalence.

By applying e), we know that, for all small $\omega > 0$, there exists $\xi \neq 0$ such that e′) holds. On the other hand – by applying d) twice – for all small $\omega > 0$, there exists $\xi \neq 0$ such that

$$(\mathbb{D}_\omega \times B_{\epsilon_2}) \cap (h + w^j)^{-1}(\mathbb{D}_\xi - 0) \hookrightarrow (\mathbb{D}_{\omega_2} \times B_{\epsilon_2}) \cap (h + w^j)^{-1}(\mathbb{D}_\xi - 0)$$

and

$$(\mathbb{D}_{\omega_2} \times B_{\epsilon_1}) \cap (h + w^j)^{-1}(\mathbb{D}_\xi - 0) \hookrightarrow (\mathbb{D}_{\omega_1} \times B_{\epsilon_1}) \cap (h + w^j)^{-1}(\mathbb{D}_\xi - 0)$$

are fibre-homotopy equivalences.

The desired conclusion follows from the two homotopy-equivalences above together with e′). □

For the final results of this appendix, we return to the setting of families of analytic functions. Again, $\mathcal{U}$ will denote an open neighborhood of the origin in $\mathbb{C}^{n+1}$ and $f_t : (\mathcal{U}, \mathbf{0}) \to (\mathbb{C}, 0)$ will be an analytic family. We continue with w being a variable disjoint from those of f_t and with $j \geqslant 2$. Recall from A.8 that $\mathfrak{T}_f$ denotes the Thom set of f at the origin.

We need the following easy lemma:

Lemma A.15. *If $V(t) \notin \mathfrak{T}_f$, then $V(t) \notin \mathfrak{T}_{f+w^j}$.*

Proof. This is completely trivial. We leave it as an exercise. □

Proposition A.16. *Suppose that $V(t) \not\in \mathfrak{T}_f$ and that the family $f_t + w^j$ has the homology Milnor fibre lifting property. Then, $\Gamma^1_{f,t} = \emptyset$ near the origin and f_t has the homology Milnor fibre lifting property.*

Moreover, if $\dim_{\mathbf{0}} \Sigma f_0 \leqslant n-2$, $V(t) \not\in \mathfrak{T}_f$, and the family $f_t + w^j$ has the homotopy Milnor fibre lifting property, then f_t has the homotopy Milnor fibration lifting property.

Proof. The second claim follows immediately from the first claim, since the condition $\dim_{\mathbf{0}} \Sigma f_0 \leqslant n-2$ implies that the Milnor fibres are simply-connected. Also, since $V(t) \not\in \mathfrak{T}_f$, we immediately have that $\Gamma^1_{f,t} = \emptyset$ near the origin. What we need to prove is that f_t has the homology Milnor fibre lifting property.

Fix a good stratification $\mathfrak{G}$ for f_0 at the origin. We must now make many choices.

1) Let $(B_{\epsilon_0}, \mathbb{D}_{\lambda_0})$ be a Milnor pair for f_0 such that
2) $B_{\epsilon_0} \cap \Sigma f_0 \subseteq V(f_0)$, and
3) ∂B_{ϵ_0} transversely intersects the strata of $\mathfrak{G}$.

From A.9, we know that the conormal condition holds, and so we may pick $\eta, \tau > 0$ such that

4) the map $G := (f, t)$ restricted to $\mathbb{C} \times \partial B_{\epsilon_0}$ has no critical values in $(\mathbb{D}_\eta - 0) \times \mathbb{D}_\tau$.

Using A.14, we may also choose $\omega_0, \xi_0 > 0$ such that

5) $(\mathbb{D}_{\omega_0} \times B_{\epsilon_0}, \mathbb{D}_{\xi_0})$ is a Milnor pair for $f_0 + w^j$, where
6) $\omega_0^j < \eta$, and
7) all of the obvious Whitney strata of $\mathbb{D}_{\omega_0} \times B_{\epsilon_0}$ transversely intersect all of the strata in the good stratification for $f_0 + w^j$ which is induced by $\mathfrak{G}$ (as given in Proposition 8.3).

Now, as $V(t) \not\in \mathfrak{T}_f$, Lemma A.15 tells us that $V(t) \not\in \mathfrak{T}_{f+w^j}$. Hence, $f_t + w^j$ satisfies the universal conormal condition and so, for all small $\nu \neq 0$ and all small t_1,

8) $(\mathbb{D}_{\omega_0} \times B_{\epsilon_0}) \cap V(f_{t_1} + w^j - \nu)$ is diffeomorphic to $F_{f_0+w^j, \mathbf{0}}$.

We select t_1 so that

9) t_1 is in $\mathbb{D}_\tau$,
10) $\Gamma^1_{f,t} \cap (\mathbb{D}_{|t_1|} \times B_{\epsilon_0}) = \emptyset$, and
11) $\Sigma f \cap (\mathbb{D}_{|t_1|} \times B_{\epsilon_0}) \subseteq V(f)$.

As t_1 is in $\mathbb{D}_\tau$, there exists λ'_0 such that

12) for all γ with $0 < \gamma < \lambda'_0$, $B_{\epsilon_0} \cap V(f_{t_1} - \gamma)$ is diffeomorphic to $F_{f_0, \mathbf{0}}$.
13) Now, let $B_\epsilon, \mathbb{D}_\lambda)$ be a Milnor pair for f_{t_1} with
14) $\epsilon < \epsilon_0$ and $\lambda < \lambda'_0$.

Then, there exist $\omega, \xi > 0$ such that

15) $(\mathbb{D}_\omega \times B_\epsilon, \mathbb{D}_\xi)$ is a Milnor pair for $f_{t_1} + w^j$, where we assume that
16) $\omega^j < \min\{\lambda, \lambda'_0, \omega_0^j\}$ and
17) $\xi < \min\{\xi_0, \omega^j, \eta - \omega_0^j\}$, where $\eta - \omega_0^j > 0$ by 6).

Finally, we select ν in 8) so small that

18) $0 < |\nu| < \min\{\eta - \omega_0^j, \lambda - \omega^j, \xi\}$.

Now that we have made all of these choices, we are ready to begin the intuitive part of the proof.

We have the inclusions

$$F_{f_{t_1}+w^j,\mathbf{0}} \cong (\mathbb{D}_\omega \times B_\epsilon) \cap V(f_{t_1} + w^j - \nu) \xrightarrow{i} (\mathbb{D}_\omega \times B_{\epsilon_0}) \cap V(f_{t_1} + w^j - \nu)$$

$$\xrightarrow{l} (\mathbb{D}_{\omega_0} \times B_{\epsilon_0}) \cap V(f_{t_1} + w^j - \nu) \cong F_{f_0+w^j,\mathbf{0}},$$

where we are assuming that $l \circ i$ induces isomorphisms on homology. We will first show that l induces isomorphisms on homology and, hence, so does i. Actually, we will show that l is a homotopy-equivalence.

We accomplish this by showing that

$$(*) \qquad \left((\mathbb{D}_{\omega_0} - \mathring{\mathbb{D}}_\omega) \times B_{\epsilon_0}\right) \cap V(f_{t_1} + w^j - \nu) \;\xrightarrow{\;w\;}\; \mathbb{D}_{\omega_0} - \mathring{\mathbb{D}}_\omega$$

is a proper, stratified submersion.

Critical points of the map in $(\mathbb{D}_{\omega_0} - \mathring{\mathbb{D}}_\omega) \times \mathring{B}_{\epsilon_0}$ occur where $\operatorname{grad}(f_{t_1}) = \mathbf{0}$; that is, at points $(w, t_1, \mathbf{z})$ such that $(t_1, \mathbf{z})$ is in $\Gamma^1_{f,t}$ or in Σf. By 10), $\Gamma^1_{f,t} \cap (\mathbb{D}_{t_1} \times B_{\epsilon_0})$ is empty and, by 11), $\Sigma f \cap (\mathbb{D}_{t_1} \times B_{\epsilon_0}) \subseteq V(f)$. But, if $f_{t_1} = 0$, then $w^j - \nu = 0$. However, this is impossible since $w \in \mathbb{D}_{\omega_0} - \mathring{\mathbb{D}}_\omega$ and thus we would have to have $|w^j| \geqslant \omega^j$ – but we know that $\omega^j > \xi > |\nu|$ by 17) and 18).

Now, we consider critical points of $(*)$ which occur on $(\mathbb{D}_{\omega_0} - \mathring{\mathbb{D}}_\omega) \times \partial B_{\epsilon_0}$. These occur at points $(w, \mathbf{p})$ where $T_\mathbf{p} V(f_{t_1} - f_{t_1}(\mathbf{p})) \subseteq T_\mathbf{p} \partial B_{\epsilon_0}$. However, $0 < |f_{t_1}(\mathbf{p})| = |w^j - \nu| \leqslant |w|^j + |\nu|$, where $0 < |w^j - \nu|$ by the argument of the preceding paragraph. But, $w \in \mathbb{D}_{\omega_0}$ and so $|w|^j + |\nu| \leqslant \omega_0^j + |\nu|$ which is $\leqslant \omega_0^j + \eta - \omega_0^j$ by 18). Hence, $0 < |f_{t_1}(\mathbf{p})| \leqslant \eta$, $t_1 \in \mathbb{D}_\tau$, and $T_\mathbf{p} V(f_{t_1} - f_{t_1}(\mathbf{p})) \subseteq T_\mathbf{p} \partial B_{\epsilon_0}$; this contradicts 4).

Therefore, the map $(*)$ is a proper, stratified submersion and, hence, is a locally trivial fibration. It follows at once that the inclusion, l, is a homotopy-equivalence and, thus, it follows that our earlier map

$$(\mathbb{D}_\omega \times B_\epsilon) \cap V(f_{t_1} + w^j - \nu) \xrightarrow{i} (\mathbb{D}_\omega \times B_{\epsilon_0}) \cap V(f_{t_1} + w^j - \nu)$$

induces isomorphisms on homology.

We wish now to show that i is obtained up to homotopy by wedging together $j-1$ copies of the suspension of the inclusion map $B_\epsilon \cap V(f_{t_1} - \nu) \hookrightarrow B_{\epsilon_0} \cap V(f_{t_1} - \nu)$ which, by 12), 13), and 18), is nothing more than the inclusion $F_{f_{t_1},\mathbf{0}} \hookrightarrow F_{f_0,\mathbf{0}}$. It would then follow that $F_{f_{t_1},\mathbf{0}} \hookrightarrow F_{f_0,\mathbf{0}}$ induces isomorphisms on homology since i does.

But, since $|w^j - \nu| \leqslant |w|^j + |\nu| \leqslant \omega^j + |\nu| \leqslant \min\{\xi, \lambda'_0\}$ by 18) and 14), we may proceed as in Proposition 8.1 and find that projection by w realizes, up to homotopy:

$(\mathbb{D}_\omega \times B_\epsilon) \cap V(f_{t_1} + w^j - \nu)$ as the wedge of $j-1$ copies of the suspension of $F_{f_{t_1},\mathbf{0}}$,

$(\mathbb{D}_\omega \times B_{\epsilon_0}) \cap V(f_{t_1} + w^j - \nu)$ as the wedge of $j-1$ copies of the suspension of $F_{f_0,\mathbf{0}}$, and

the map i as the wedge of $j-1$ copies of the suspension of the map

$$F_{f_{t_1},\mathbf{0}} \cong B_\epsilon \cap V(f_{t_1} - \nu) \hookrightarrow B_{\epsilon_0} \cap V(f_{t_1} - \nu) \cong F_{f_0,\mathbf{0}}.$$

The conclusion follows. □

References

[BBD] A. Beilinson, J. Berstein, and P. Deligne, *Faisceaux Pervers*, Astérisque **100**, Soc. Math. de France, 1983.

[B-S] J. Briançon and J.P. Speder, *La trivialité topologique n'implique pas les conditions de Whitney*, C.R. Acad. Sci. Paris, Série A **280** (1975).

[Br] J. Brylinski, *Transformations canoniques, Dualité projective, Théorie de Lefschetz, Transformations de Fourier et sommes trigonométriques*, Soc. Math. de France, Astérisque **140** (1986).

[BDK] J. Brylinski, A. Dubson, and M. Kashiwara, *Formule de l'indice pour les modules holonomes et obstruction d'Euler locale*, C.R. Acad. Sci., Série A **293** (1981), 573–576.

[Ce] J. Cerf, *La stratification naturelle des espaces de fonctions différentiables réelles et le théorème de la pseudo-isotopie*, publ. I.H.E.S. **39** (1970), 187–353.

[Co] D. Cohen, *unpublished note*, 1991.

[De] P. Deligne, *Comparaison avec la théorie transcendante*, Séminaire de géométrie algébrique du Bois-Marie, SGA 7 II, Springer Lect. Notes **340** (1973).

[Di] A. Dimca, *On the Milnor fibration of weighted homogeneous polynomials*, Compositio Math. **76** (1990), 19–47.

[Fu] W. Fulton, *Intersection Theory*, Ergebnisse der Math., Springer-Verlag, 1984.

[Gaf1] T. Gaffney, *personal communication*, 1991.

[Gaf2] ———, *Polar Multiplicities and Equisingularity of Map Germs*, Topology (to appear).

[Gas1] L. van Gastel, *Excess Intersections*, Thesis, University of Utrecht, 1989.

[Gas2] ———, *Excess Intersections and a Correspondence Principle*, Invent. Math. **103 (1)** (1991), 197–222.

[Gi] V. Ginsburg, *Characteristic Varieties and Vanishing Cycles*, Invent. Math. **84** (1986), 327–403.

[G-M1] M. Goresky and R. MacPherson, *Morse Theory and Intersection Homology*, Analyse et Topologie sur les Espaces Singuliers. Astérisque **101** (1983), Soc. Math. France, 135–192.

[G-M2] ———, *Stratified Morse Theory*, Ergebnisse der Math. 14, Springer-Verlag, Berlin, 1988.

[G-M3] ———, *Intersection homology II*, Inv. Math **71** (1983), 77–129.

[G-M4] ———, *Intersection homology theory*, Topology **19** (1980), 135–162.

[G-R1] H. Grauert and R. Remmert, *Coherent Analytic Sheaves*, Grund. math. Wiss. 265, Springer-Verlag, 1984.

[G-R2] ———, *Theory of Stein Spaces*, Grund. math. Wiss. 236, Springer-Verlag, 1979.

[Gr] G. M. Greuel, *Constant Milnor number implies constant multiplicity for quasihomogeneous singularities*, Manuscr. Math. **56** (1986), 159–166.

[H-L] H. Hamm and Lê D. T., *Un Théorème de Zariski du type de Lefschetz*, Ann. Sci. L'Ecole Norm. Sup. **6** (1973), 317–366.

[Ha] R. Hartshorne, *Residues and Duality*, Springer Lecture Notes 20, Springer-Verlag, 1966.

[H-M] J.-P. Henry and M. Merle, *Conditions de régularité et éclatements*, Ann. Inst. Fourier **37** (1987), 159–190.

[HMS] J.-P. Henry, M. Merle, and C. Sabbah, *Sur la condition de Thom stricte pour un morphisme analytique*, Ann. Sci. L'Ecole Norm. Sup. **17** (1984), 227–268.

[Hi] H. Hironaka, *Normal Cones in Analytic Whitney Stratifications*, I.H.E.S. **36** (1970), 127–138.

[Hu] D. Husemoller, *Fibre Bundles*, Grad. Text in Math. 20, Springer-Verlag, 1966.

[Io] I. N. Iomdin, *Variétés complexes avec singularités de dimension un*, Sibirsk. Mat. Z. **15** (1974), 1061–1082.

[Iv] B. Iverson, *Cohomology of Sheaves*, Universitext, Springer-Verlag, 1986.

[K-S1] M. Kashiwara and P. Schapira, *Microlocal Study of Sheaves*, Astérisque **128** (1985).

[K-S2] M. Kashiwara and P. Schapira, *Sheaves on Manifolds*, Grund. math. Wiss. 292, Springer - Verlag, 1990.

[K-M] M. Kato and Y. Matsumoto, *On the connectivity of the Milnor fibre of a holomorphic function at a critical point*, Proc. of 1973 Tokyo manifolds conf. (1973), 131–136.

[La] R. Lazarsfeld, *Branched Coverings of Projective Space*, Thesis, Brown University, 1980.

[Lê1] Lê D. T., *Calcul du Nombre de Cycles Évanouissants d'une Hypersurface Complexe*, Ann. Inst. Fourier, Grenoble **23** (1973), 261–270.

[Lê2] ______, *Ensembles analytiques complexes avec lieu singulier de dimension un (d'après I.N. Iomdin)*, Séminaire sur les Singularités (Paris, 1976–1977) Publ. Math. Univ. Paris VII (1980), 87–95.

[Lê3] ______, *La Monodromie n'a pas de Points Fixes*, J. Fac. Sci. Univ. Tokyo, Sec. 1A **22** (1975), 409–427.

[Lê4] ______, *Morsification of $\mathcal{D}$-Modules*, preprint (1988).

[Lê5] ______, *Some Remarks on Relative Monodromy*, Real and Complex Singularities, Oslo 1976 (1977), 397–403.

[Lê6] ______, *Sur les cycles évanouissants des espaces analytiques*, C.R. Acad. Sci. Paris, Ser. A **288** (1979), 283–285.

[Lê7] ______, *Topologie des Singularités des Hypersurfaces Complexes*, Astérisque **7** and **8** (1973), 171–192.

[Lê8] ______, *Une application d'un théorème d'A'Campo a l'equisingularité*, Indag. Math **35** (1973), 403–409.

[L-M] Lê D. T. and Z. Mebkhout, *Variétés caractéristiques et variétés polaires*, C.R. Acad. Sci. **296** (1983), 129–132.

[L-P] Lê D. T. and B. Perron, *Sur la Fibre de Milnor d'une Singularité Isolée en Dimension Complexe Trois*, C.R. Acad. Sci. **289** (1979), 115-118.

[L-R] Lê D. T. and C. P. Ramanujam, *The Invariance of Milnor's Number implies the Invariance of the Topological Type*, Amer. Journ. Math. **98** (1976), 67–78.

[L-S] Lê D. T. and K. Saito, *La constance du nombre de Milnor donne des bonnes stratifications*, C.R. Acad. Sci. **277** (1973), 793–795.

[L-T1] Lê D. T. and B. Teissier, *Cycles evanescents, sections planes et conditions de Whitney. II*, Proc. Symp. Pure Math. **40, Part 2** (1983), 65–103.

[L-T2] ———, *Variétés polaires locales et classes de Chern des variétiés singulières*, Annals of Math. **114** (1981), 457–491.

[Lo] E. Looijenga, *Isolated singular points on complete intersections*, London Math. Soc. Lect. Note Series, no. 77, 1984.

[Mac1] R. MacPherson, *Chern classes for singular algebraic varieties*, Annals of Math. **100** (1974), 423–432.

[Mac2] ———, *Global Questions in the Topology of Singular Spaces*, Proc. Internat. Congress of Math., Warsaw (1983), 213–235.

[Mas1] D. Massey, *Families Of Hypersurfaces with One-Dimensional Singular Sets*, Dissertation, Duke University (1986).

[Mas2] ———, *The Lê-Ramanujam Problem for Hypersurfaces with One-Dimensional Singular Sets*, Math. Annalen **282** (1988), 33–49.

[Mas3] ———, *The Lê Varieties, I*, Invent. Math. **99** (1990), 357–376.

[Mas4] ———, *The Lê Varieties, II*, Invent. Math. **104** (1991), 113–148.

[Mas5] ———, *Numerical Invariants of Perverse Sheaves*, Duke Math. J. **73** (1994), 307–369.

[Mas6] ———, *The Thom Condition along a Line*, Duke Math. J. **60** (1990), 631–642.

[Mas7] ———, *A Reduction Theorem for the Zariski Multiplicity Conjecture*, Proc. AMS **106** (1989), 379–383.

[M-S] D. Massey and D. Siersma, *Deformations of Polar Methods*, Ann. Inst. Fourier **42** (1992), 737–778.

[MSSVWZ] D. Massey, R. Simion, R. Stanley, D. Vertigan, D. Welsh, and G. Ziegler, *Lê Numbers, Matroid Identities, and the Tutte Polynomial*, preprint, 1994.

[Mat] J. Mather, *Notes on Topological Stability*, unpublished notes, Harvard Univ. 1970.

[Me] Z. Mebkhout, *Local cohomology of analytic spaces*, Pub. Res. Inst. Math. Sc. **12** (1977), 247–256.

[Mi1] J. Milnor, *Lectures on the h-cobordism Theorem*, Math. Notes 1, P.U.P., 1965.

[Mi2] ———, *Morse Theory*, Annals of Math. Studies, no. 51, P.U.P., 1963.

[Mi3] ———, *Singular Points of Complex Hypersurfaces*, Annals of Math. Studies, no. 77, P.U.P., 1968.

[M-O] J. Milnor and P. Orlik, *Isolated Singularities Defined by Weighted Homogeneous Polynomials*, Topology **9** (1969), 385–393.

[Ok] M. Oka, *On the homotopy type of hypersurfaces defined by weighted homogeneous polynomials*, Topology **12** (1973), 19–32.

[O-R] P. Orlik and R. Randell, *The Milnor fiber of a generic arrangement*, Arkiv für Mat. **31** (1993), 71–81.

[O-S] P. Orlik and L. Solomon, *Combinatorics and topology of complements of hyperplanes*, Invent. Math. **56** (1980), 167–189.

[O-T] P. Orlik and H. Terao, *Arrangements of Hyperplanes*, Grund. math. Wiss., vol. 300, Springer-Verlag, 1991.

[O'S] D. O'Shea, *Topologically Trivial Deformations of Isolated Quasihomogeneous Hypersurface Singularities are Equimultiple*, Proc. AMS **100** (1987), 260–262.

[Ra] R. Randell, *On the Topology of Non-isolated Singularities*, Proc. Georgia Top. Conf., Athens, Ga., 1977 **99** (1979), 445–473.

[Sab] C. Sabbah, *Quelques remarques sur la géométrie des espaces conormaux*, Astérisque **130** (1985), 161–192.

[Sak] K. Sakamoto, *The Seifert matrices of Milnor fiberings defined by holomorphic functions*, J. Math. Soc. Japan **26 (4)** (1974), 714–721.

[Sc-To] P. Schapira and N. Tose, *Morse Inequalities for R-Constructible Sheaves*, Adv. in Math. **93** (1992), 1–8.

[Se-Th] M. Sebastiani and R. Thom, *Un résultat sur la monodromie*, Invent. Math. **13** (1971), 90–96.

[Si1] D. Siersma, *Isolated Line Singularities*, Proc. Symp. Pure Math. **40, Part 2** (1983), 485–496.

[Si2] ———, *The monodromy of a series of hypersurface singularities*, Comment. Math. Helvetici **65** (1990), 181–197.

[Si-Tr] Y. T. Siu and G. Trautmann, *Gap-Sheaves and Extension of Coherent Analytic Subsheaves*, Springer Lect. Notes **172**, Springer-Verlag, 1971.

[Sm] S. Smale, *Generalized Poincaré's Conjecture in Dimensions greater than 4*, Ann. Math. **64** (1956).

[Te1] B. Teissier, *The Hunting of Invariants in the Geometry of Discriminants*, in Real and Complex Singularities, Oslo 1976, ed. P. Holm (1977), 565–677.

[Te2] ———, *Introduction to Equisingularity Problems*, Proc. Symp. Pure Math. **29** (1975), 593–632.

[Te3] ———, *Variétés polaires I: Invariants polaires des singularités d'hypersurfaces*, Invent. Math. **40 (3)** (1977), 267–292.

[Te4] ———, *Variétés polaires II: Multiplicités polaires, sections planes, et conditions de Whitney*, in Algebraic Geometry, Proc., La Rabida 1981, Springer Lect. Notes **961** (1982), 314–491.

[Th] R. Thom, *Ensembles et Morphismes Stratifiés*, Bull. Amer. Math. Soc. **75** (1969), 240–284.

[Ti] M. Tibăr, *Bouquet Decomposition of the Milnor Fibre* (to appear).

[Va1] J. P. Vannier, *Familles à paramètre de fonctions holomorphes à ensemble singulier de dimension zéro ou un*, Thèse, Dijon (1987).

[Va2] ———, *Sur les fibrations de Milnor de familles d'hypersurfaces à lieu singulier de dimension un*, Math. Ann. **287** (1990), 539–552.

[Vo] W. Vogel, *Results on Bézout's Theorem*, Tata Lecture Notes 74, Springer-Verlag, 1984.

[W] H. Whitney, *Tangents to an Analytic Variety*, Ann. Math. **81** (1965), 496–549.

[Z] O. Zariski, *Open Questions in the Theory of Singularities*, Bull. AMS **77** (1971), 481–491.

SUBJECT INDEX

Lecture Notes in Mathematics

For information about Vols. 1–1439
please contact your bookseller or Springer-Verlag

Vol. 1440: P. Latiolais (Ed.), Topology and Combinatorial Groups Theory. Seminar, 1985–1988. VI, 207 pages. 1990.

Vol. 1441: M. Coornaert, T. Delzant, A. Papadopoulos. Géométrie et théorie des groupes. X, 165 pages. 1990.

Vol. 1442: L. Accardi, M. von Waldenfels (Eds.), Quantum Probability and Applications V. Proceedings, 1988. VI, 413 pages. 1990.

Vol. 1443: K.H. Dovermann, R. Schultz, Equivariant Surgery Theories and Their Periodicity Properties. VI, 227 pages. 1990.

Vol. 1444: H. Korezlioglu, A.S. Ustunel (Eds.), Stochastic Analysis and Related Topics VI. Proceedings, 1988. V, 268 pages. 1990.

Vol. 1445: F. Schulz, Regularity Theory for Quasilinear Elliptic Systems and – Monge Ampère Equations in Two Dimensions. XV, 123 pages. 1990.

Vol. 1446: Methods of Nonconvex Analysis. Seminar, 1989. Editor: A. Cellina. V, 206 pages. 1990.

Vol. 1447: J.-G. Labesse, J. Schwermer (Eds), Cohomology of Arithmetic Groups and Automorphic Forms. Proceedings, 1989. V, 358 pages. 1990.

Vol. 1448: S.K. Jain, S.R. López-Permouth (Eds.), Non-Commutative Ring Theory. Proceedings, 1989. V, 166 pages. 1990.

Vol. 1449: W. Odyniec, G. Lewicki, Minimal Projections in Banach Spaces. VIII, 168 pages. 1990.

Vol. 1450: H. Fujita, T. Ikebe, S.T. Kuroda (Eds.), Functional-Analytic Methods for Partial Differential Equations. Proceedings, 1989. VII, 252 pages. 1990.

Vol. 1451: L. Alvarez-Gaumé, E. Arbarello, C. De Concini, N.J. Hitchin, Global Geometry and Mathematical Physics. Montecatini Terme 1988. Seminar. Editors: M. Francaviglia, F. Gherardelli. IX, 197 pages. 1990.

Vol. 1452: E. Hlawka, R.F. Tichy (Eds.), Number-Theoretic Analysis. Seminar, 1988–89. V, 220 pages. 1990.

Vol. 1453: Yu.G. Borisovich, Yu.E. Gliklikh (Eds.), Global Analysis – Studies and Applications IV. V, 320 pages. 1990.

Vol. 1454: F. Baldassari, S. Bosch, B. Dwork (Eds.), p-adic Analysis. Proceedings, 1989. V, 382 pages. 1990.

Vol. 1455: J.-P. Françoise, R. Roussarie (Eds.), Bifurcations of Planar Vector Fields. Proceedings, 1989. VI, 396 pages. 1990.

Vol. 1456: L.G. Kovács (Ed.), Groups – Canberra 1989. Proceedings. XII, 198 pages. 1990.

Vol. 1457: O. Axelsson, L.Yu. Kolotilina (Eds.). Preconditioned Conjugate Gradient Methods. Proceedings, 1989. V, 196 pages. 1990.

Vol. 1458: R. Schaaf, Global Solution Branches of Two Point Boundary Value Problems. XIX, 141 pages. 1990.

Vol. 1459: D. Tiba, Optimal Control of Nonsmooth Distributed Parameter Systems. VII, 159 pages. 1990.

Vol. 1460: G. Toscani, V. Boffi, S. Rionero (Eds.), Mathematical Aspects of Fluid Plasma Dynamics. Proceedings, 1988. V, 221 pages. 1991.

Vol. 1461: R. Gorenflo, S. Vessella, Abel Integral Equations. VII, 215 pages. 1991.

Vol. 1462: D. Mond, J. Montaldi (Eds.), Singularity Theory and its Applications. Warwick 1989, Part I. VIII, 405 pages. 1991.

Vol. 1463: R. Roberts, I. Stewart (Eds.), Singularity Theory and its Applications. Warwick 1989, Part II. VIII, 322 pages. 1991.

Vol. 1464: D. L. Burkholder, E. Pardoux, A. Sznitman, Ecole d'Eté de Probabilités de Saint- Flour XIX-1989. Editor: P. L. Hennequin. VI, 256 pages. 1991.

Vol. 1465: G. David, Wavelets and Singular Integrals on Curves and Surfaces. X, 107 pages. 1991.

Vol. 1466: W. Banaszczyk, Additive Subgroups of Topological Vector Spaces. VII, 178 pages. 1991.

Vol. 1467: W. M. Schmidt, Diophantine Approximations and Diophantine Equations. VIII, 217 pages. 1991.

Vol. 1468: J. Noguchi, T. Ohsawa (Eds.), Prospects in Complex Geometry. Proceedings, 1989. VII, 421 pages. 1991.

Vol. 1469: J. Lindenstrauss, V. D. Milman (Eds.), Geometric Aspects of Functional Analysis. Seminar 1989-90. XI, 191 pages. 1991.

Vol. 1470: E. Odell, H. Rosenthal (Eds.), Functional Analysis. Proceedings, 1987-89. VII, 199 pages. 1991.

Vol. 1471: A. A. Panchishkin, Non-Archimedean L-Functions of Siegel and Hilbert Modular Forms. VII, 157 pages. 1991.

Vol. 1472: T. T. Nielsen, Bose Algebras: The Complex and Real Wave Representations. V, 132 pages. 1991.

Vol. 1473: Y. Hino, S. Murakami, T. Naito, Functional Differential Equations with Infinite Delay. X, 317 pages. 1991.

Vol. 1474: S. Jackowski, B. Oliver, K. Pawałowski (Eds.), Algebraic Topology, Poznań 1989. Proceedings. VIII, 397 pages. 1991.

Vol. 1475: S. Busenberg, M. Martelli (Eds.), Delay Differential Equations and Dynamical Systems. Proceedings, 1990. VIII, 249 pages. 1991.

Vol. 1476: M. Bekkali, Topics in Set Theory. VII, 120 pages. 1991.

Vol. 1477: R. Jajte, Strong Limit Theorems in Noncommutative L_2-Spaces. X, 113 pages. 1991.

Vol. 1478: M.-P. Malliavin (Ed.), Topics in Invariant Theory. Seminar 1989-1990. VI, 272 pages. 1991.

Vol. 1479: S. Bloch, I. Dolgachev, W. Fulton (Eds.), Algebraic Geometry. Proceedings, 1989. VII, 300 pages. 1991.

Vol. 1480: F. Dumortier, R. Roussarie, J. Sotomayor, H. Żołądek, Bifurcations of Planar Vector Fields: Nilpotent Singularities and Abelian Integrals. VIII, 226 pages. 1991.

Vol. 1481: D. Ferus, U. Pinkall, U. Simon, B. Wegner (Eds.), Global Differential Geometry and Global Analysis. Proceedings, 1991. VIII, 283 pages. 1991.

Vol. 1482: J. Chabrowski, The Dirichlet Problem with L^2-Boundary Data for Elliptic Linear Equations. VI, 173 pages. 1991.

Vol. 1483: E. Reithmeier, Periodic Solutions of Nonlinear Dynamical Systems. VI, 171 pages. 1991.

Vol. 1484: H. Delfs, Homology of Locally Semialgebraic Spaces. IX, 136 pages. 1991.

Vol. 1485: J. Azéma, P. A. Meyer, M. Yor (Eds.), Séminaire de Probabilités XXV. VIII, 440 pages. 1991.

Vol. 1486: L. Arnold, H. Crauel, J.-P. Eckmann (Eds.), Lyapunov Exponents. Proceedings, 1990. VIII, 365 pages. 1991.

Vol. 1487: E. Freitag, Singular Modular Forms and Theta Relations. VI, 172 pages. 1991.

Vol. 1488: A. Carboni, M. C. Pedicchio, G. Rosolini (Eds.), Category Theory. Proceedings, 1990. VII, 494 pages. 1991.

Vol. 1489: A. Mielke, Hamiltonian and Lagrangian Flows on Center Manifolds. X, 140 pages. 1991.

Vol. 1490: K. Metsch, Linear Spaces with Few Lines. XIII, 196 pages. 1991.

Vol. 1491: E. Lluis-Puebla, J.-L. Loday, H. Gillet, C. Soulé, V. Snaith, Higher Algebraic K-Theory: an overview. IX, 164 pages. 1992.

Vol. 1492: K. R. Wicks, Fractals and Hyperspaces. VIII, 168 pages. 1991.

Vol. 1493: E. Benoît (Ed.), Dynamic Bifurcations. Proceedings, Luminy 1990. VII, 219 pages. 1991.

Vol. 1494: M.-T. Cheng, X.-W. Zhou, D.-G. Deng (Eds.), Harmonic Analysis. Proceedings, 1988. IX, 226 pages. 1991.

Vol. 1495: J. M. Bony, G. Grubb, L. Hörmander, H. Komatsu, J. Sjöstrand, Microlocal Analysis and Applications. Montecatini Terme, 1989. Editors: L. Cattabriga, L. Rodino. VII, 349 pages. 1991.

Vol. 1496: C. Foias, B. Francis, J. W. Helton, H. Kwakernaak, J. B. Pearson, H_∞-Control Theory. Como, 1990. Editors: E. Mosca, L. Pandolfi. VII, 336 pages. 1991.

Vol. 1497: G. T. Herman, A. K. Louis, F. Natterer (Eds.), Mathematical Methods in Tomography. Proceedings 1990. X, 268 pages. 1991.

Vol. 1498: R. Lang, Spectral Theory of Random Schrödinger Operators. X, 125 pages. 1991.

Vol. 1499: K. Taira, Boundary Value Problems and Markov Processes. IX, 132 pages. 1991.

Vol. 1500: J.-P. Serre, Lie Algebras and Lie Groups. VII, 168 pages. 1992.

Vol. 1501: A. De Masi, E. Presutti, Mathematical Methods for Hydrodynamic Limits. IX, 196 pages. 1991.

Vol. 1502: C. Simpson, Asymptotic Behavior of Monodromy. V, 139 pages. 1991.

Vol. 1503: S. Shokranian, The Selberg-Arthur Trace Formula (Lectures by J. Arthur). VII, 97 pages. 1991.

Vol. 1504: J. Cheeger, M. Gromov, C. Okonek, P. Pansu, Geometric Topology: Recent Developments. Editors: P. de Bartolomeis, F. Tricerri. VII, 197 pages. 1991.

Vol. 1505: K. Kajitani, T. Nishitani, The Hyperbolic Cauchy Problem. VII, 168 pages. 1991.

Vol. 1506: A. Buium, Differential Algebraic Groups of Finite Dimension. XV, 145 pages. 1992.

Vol. 1507: K. Hulek, T. Peternell, M. Schneider, F.-O. Schreyer (Eds.), Complex Algebraic Varieties. Proceedings, 1990. VII, 179 pages. 1992.

Vol. 1508: M. Vuorinen (Ed.), Quasiconformal Space Mappings. A Collection of Surveys 1960-1990. IX, 148 pages. 1992.

Vol. 1509: J. Aguadé, M. Castellet, F. R. Cohen (Eds.), Algebraic Topology - Homotopy and Group Cohomology. Proceedings, 1990. X, 330 pages. 1992.

Vol. 1510: P. P. Kulish (Ed.), Quantum Groups. Proceedings, 1990. XII, 398 pages. 1992.

Vol. 1511: B. S. Yadav, D. Singh (Eds.), Functional Analysis and Operator Theory. Proceedings, 1990. VIII, 223 pages. 1992.

Vol. 1512: L. M. Adleman, M.-D. A. Huang, Primality Testing and Abelian Varieties Over Finite Fields. VII, 142 pages. 1992.

Vol. 1513: L. S. Block, W. A. Coppel, Dynamics in One Dimension. VIII, 249 pages. 1992.

Vol. 1514: U. Krengel, K. Richter, V. Warstat (Eds.), Ergodic Theory and Related Topics III, Proceedings, 1990. VIII, 236 pages. 1992.

Vol. 1515: E. Ballico, F. Catanese, C. Ciliberto (Eds.), Classification of Irregular Varieties. Proceedings, 1990. VII, 149 pages. 1992.

Vol. 1516: R. A. Lorentz, Multivariate Birkhoff Interpolation. IX, 192 pages. 1992.

Vol. 1517: K. Keimel, W. Roth, Ordered Cones and Approximation. VI, 134 pages. 1992.

Vol. 1518: H. Stichtenoth, M. A. Tsfasman (Eds.), Coding Theory and Algebraic Geometry. Proceedings, 1991. VIII, 223 pages. 1992.

Vol. 1519: M. W. Short, The Primitive Soluble Permutation Groups of Degree less than 256. IX, 145 pages. 1992.

Vol. 1520: Yu. G. Borisovich, Yu. E. Gliklikh (Eds.), Global Analysis – Studies and Applications V. VII, 284 pages. 1992.

Vol. 1521: S. Busenberg, B. Forte, H. K. Kuiken, Mathematical Modelling of Industrial Process. Bari, 1990. Editors: V. Capasso, A. Fasano. VII, 162 pages. 1992.

Vol. 1522: J.-M. Delort, F. B. I. Transformation. VII, 101 pages. 1992.

Vol. 1523: W. Xue, Rings with Morita Duality. X, 168 pages. 1992.

Vol. 1524: M. Coste, L. Mahé, M.-F. Roy (Eds.), Real Algebraic Geometry. Proceedings, 1991. VIII, 418 pages. 1992.

Vol. 1525: C. Casacuberta, M. Castellet (Eds.), Mathematical Research Today and Tomorrow. VII, 112 pages. 1992.